The Bankside Book of Stillwater Trout Flies

For Jennifer. Without her patience and understanding this book could not have been written.

The Bankside Book of Stillwater Trout Flies

Peter Lapsley

Adam and Charles Black · London

First published 1978
by A. & C. Black Ltd.
35 Bedford Row, London WC1R 4JH

© 1978 Peter Lapsley

Lapsley, Peter
  The bankside book of stillwater trout flies.
  1. Trout fishing   2. Flies, Artificial
  I. Title
  799.1′7′55   SH451

ISBN 0-7136-1894-9

Typeset in Great Britain by Eta Services (Typesetters) Ltd., Beccles, Suffolk
Printed in Great Britain by Partridge & Love, Bristol

Contents

Illustrations

Colour plates

(between pages 16 and 17)

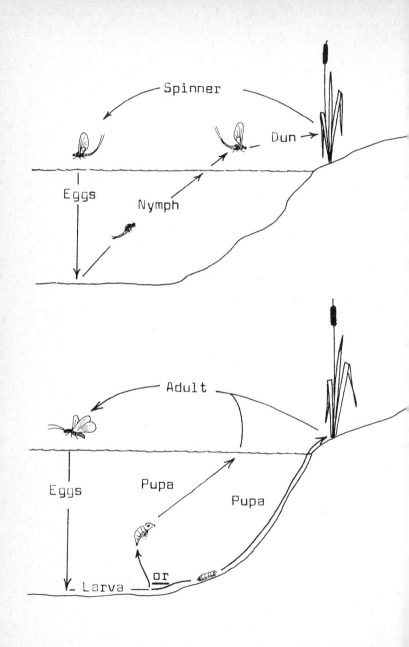

Spinner

Dun →

Eggs

Nymph

Adult

Eggs

Pupa

Pupa

or

Larva

Introduction

The problem of fly selection has taxed the minds of trout fishermen since the sport first started. Confronted with a box of traditional or fancy patterns, most anglers indulge in a form of lucky dip, perhaps tempering random choice with one of the rules of thumb such as 'bright day, bright fly; dull day, dull fly', or with some other less scientific guideline drawn from personal experience or local folklore.

Although thousands of trout are caught on lures and traditional or attractor patterns each year, the proportion of stillwater flyfishermen who prefer to use dressings specifically designed to imitate or suggest natural insects, snails, fry and so on is increasing steadily.

The lucky dip system of fly selection rarely works when applied to imitative patterns. The reasons are obvious. There is far more to imitative fishing than simply choosing a fly, casting it out and retrieving it in the hope that a fish may take hold. The selection must be made in relation to the naturals in the water upon which the fish are (or may be expected to be) feeding at the time, and the chosen pattern must resemble its natural counterpart reasonably closely in terms of size, colour and shape. (Interestingly, too exact imitation can be counter-productive and it generally pays to use an artificial which is a caricature or exaggeration of the real thing, particularly in size. Similarly, there is often advantage in incorporating materials, especially furs, into the dressings to provide translucency, sheen and movement. Such materials may reduce the

---

Left: Life cycles
Top: Upwinged or day-flies –Caenis, Claret dun, Lake Olive, Pond Olive, Mayfly, Sepia dun, Blue Winged Olive.
Bottom: Most other insects—midges, sedges, the Alder, Damselflies, Stoneflies.

realism of the artificials from the human viewpoint, but more than compensate for this by imparting 'life' to them underwater.)

Having selected the artificial, it must be presented to the fish realistically; that is to say, it must be made to behave in the the same way as its living equivalent and it should be fished in an area in which the naturals are to be found and in which trout may be expected to be feeding on them.

The initial selection of an imitative pattern may be made in any one of several ways. Obviously, the fisherman may, if he is very lucky, see trout feeding on an identifiable species and choose a fly to match the natural. Or he may notice that large numbers of a particular kind of insect are in evidence in, on, or above the water and select his artificial accordingly. Failing this, he may know that at a particular time, in a particular place, insects of a certain species are likely to be in evidence and that the trout habitually feed on them; armed with this knowledge, he is well placed to pick from his fly box a pattern which stands a good chance of proving acceptable to the fish. Finally, having caught a trout, he is in a position to examine its stomach contents and to select a pattern to match his findings. But, of course, before he can use this last method the fisherman must have caught a fish, and to this end one of the earlier methods of selection will have to be used, or a general 'food suggesting' fly may have to be tried.

Fortunately, Mother Nature is a reasonably predictable old lady. Trout diets are much the same from Cornwall to Kent and from Sussex to Scotland. Each species of animal minutiae upon which the fish feed is similar in appearance wherever it is found and, within certain broad margins, the various insects are available as potential fish food at roughly the same times of year throughout Britain. This being so, it is quite possible to select those insects and other creatures that make up the bulk of the diet of most trout in stillwaters (there are only about 16 species in all), and to describe their appearances, habits and seasons in the knowledge that most of the information will be as valid in Scotland as in southern England. Starting with this premise, my aim has been to produce a simple companion guide to what trout eat in stillwaters, and when. I hope that, armed with it, the lake or reservoir fisherman may be able to decide which creatures the trout are likely to be feeding on at a particular time of day and at a specific stage in the season, and to select an appropriate artificial pattern with which to confront the fish. The text describes the appearance, life cycle and habits of each creature, important factors if the angler is to present his pattern in as lifelike a manner as possible. Professionally dressed flies to represent the various insects are available from tackle shops. For those who wish to tie their own, each section contains dressings for tried and trusted patterns.

Because I am convinced that careful and thoughtful presentation is at least as important as having a good imitation at the end of the leader, I have included a separate set of general thoughts on this subject.

Having said all this, two notes of caution should be sounded. Some insects are relatively localised and are only found in certain areas, so the guide should not be treated as a bible for any particular fishery. If the reader is fishing on unfamiliar ground he should temper his use of the book with informed local advice. He should also be aware that irregularities in the climate can produce similar irregularities in the insects' development programmes. If the winter and spring have been cold, some insects may hatch as much as a fortnight or three weeks late. In extreme circumstances, a particular species may almost vanish for a complete year. During the long, very hot summer of 1976 there were masses of midges at my home stillwater fishery but sedge hatches were fairly sparse. In 1977, a cold, wet, blustery summer, we saw relatively few of the normally abundant midges but fished among clouds of sedges from the end of June onwards.

This slender volume does not purport to be a definitive or comprehensive work on stillwater entomology. Indeed, I am unqualified to produce such a work and have been at pains to look at the various creatures described through layman's eyes and to keep technical jargon to an absolute minimum. It has been contended, and I agree, that observation is the single most important factor governing success in fishing. I would be sorry if anyone tried to treat this little book as a substitute for intelligent observation of the insect and other life around them at the waterside. It has been written in the hope that newcomers to our sport, and those fly fishermen who still depend exclusively upon lures and traditional or attractor patterns, may venture into the imitative field with some basic information available to them, and thus with some confidence.

Finally, although this guide is exclusively concerned with so-called imitative fly-fishing, I believe that we should beware of the 'imitative snobbery' which seems to be growing around our reservoirs and lakes. Imitative fishing panders to the trout's predominant and most constant instinct — the feeding one; its exponents require a degree of knowledge if they are to be consistently successful, and it provides them with a stimulating mental challenge. Once mastered, it can often be dramatically effective. But it is only one branch of the sport and should be thought of as no more than a single weapon in an armoury of assorted methods. What does the exclusively imitative fisherman do if the fish are simply not feeding? The truly skilful angler can tailor his technique to the needs of the day.

On presentation

Traditional or attractor patterns generally rely on

arousing the trout's curiosity through their colour, brilliance and the often relatively high speeds at which people retrieve them. The natural creatures upon which trout prey are rarely bright, or colourful, nor are they fast moving. For these reasons, artificial patterns specifically dressed to represent natural insects, snails or fry make pretty poor attractors. Obviously, imitative flies must not only look like their living counterparts, they must also behave like them if they are to catch fish.

Imitative artificials must arrive unobtrusively (very few real midge pupae or Lake Olive nymphs plummet into the water from the sky) and in order to achieve this, a double tapered line is generally better than a weight forward one or a standard shooting head. While it is true that one cannot cast as long a line with a double taper, a small sacrifice of distance is well repaid if the fly arrives gently on the water.

The pundits will still be arguing about the relative merits of various line colours when there are no trout left to fish for. I am disinclined to take issue with nature; the heron, a far better fisherman than I shall ever be, seems to get by with a light coloured belly and I follow his example when selecting a floating line, even though it must be seen by the trout in silhouette. For sinking lines, likely to be seen against a background of mud, gravel or aquatic vegetation, I am happy to use dark green, dark brown or even black. More important than colour, perhaps, is line flash, and I look forward to the day when a truly matt finished line becomes available.

The line's job is to carry the leader, and the fly attached to it, out over the water. The leader is probably the most important single item of the fisherman's tackle. It must be as nearly invisible as possible, it must be capable of presenting the fly accurately and of turning over to fall in a straight line on the surface, and it must be strong enough to hold the charging run of an unexpectedly heavy fish. There is currently a vogue for leaders of 18 ft or more in length. While they unquestionably keep the fly well removed from the bulky and therefore potentially fish-frightening fly line, I do not believe that they are necessary or even suitable for fly patterns fished close to the surface. Because they inevitably contain long sections of level monofilament, they are difficult to cast with any real accuracy, particularly in a wind, and it is almost impossible to put them down on the water in a straight line. The leader that I use for fishing close to the surface or in shallow water consists of 18–24 in. of 18½ lb breaking

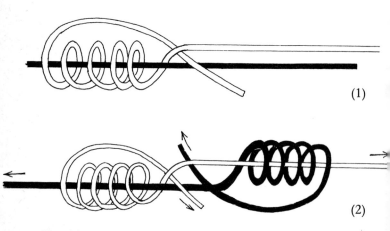

Blood knot. (1) Take two lengths of nylon of approximately equal thicknesses and overlap their ends by about 4 in. Take at least four turns of one around the other and then pass the end back between the two main strands.

(2) Repeat the process with the end of the second length, ensuring that its loose end passes through the gap between the coils in opposition to the other loose end. Moisten the knot so that the nylon will slip against itself and gently pull tight.

strain nylon monofilament, needle knotted to my fly line, then a standard, nine ft, knotless, tapered leader (generally 2× or 3×) blood knotted to this butt length, and finally, two or three ft of four or five lb breaking strain monofilament blood knotted to the point of the tapered leader. Tapered leaders are fairly expensive but this formula provides me with an excellent, almost knot-free length of nylon which meets all the requirements set out above. When I wish to use a weighted fly with a floating line, I simply replace the short point with one of perhaps five or six ft and I can then fish at any depth up to about eight or ten ft — beyond these depths I use a sinking line. With this

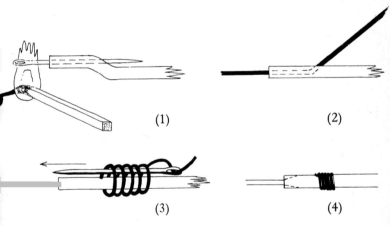

(1)   (2)

(3)   (4)

Needle knot. (1) Insert needle $\frac{1}{4}$ in. into core of fly line and out through side; heat needle briefly and withdraw.
(2) Thread monofilament into needle hole in fly line and out through side.
(3) Take at least four turns of monofilament around line and needle; pull loose end through coils with needle.
(4) Snug monofilament coils down close together and pull both ends of monofilament tight. Cut waste end of nylon away. Trim point of fly line.

longer leader, the accuracy and straightness of my casting become markedly worse but this rarely matters as, when working the fly deep down, I am almost invariably fishing the water, and the leader will straighten as the fly sinks. The needle knot between the fly line and the monofilament butt length is essential. Any other means of joining the two is sure to be bulky and cause line wake on the surface and, with any leader of more than nine ft, it is important that the line to leader knot should be able to pass freely back and forth through the rod rings.

There is now a knotless, tapered, five metre leader on the market. I have only been using it for one season but am already convinced that it will prove to be the answer to the imitative fisherman's prayers.

It is vital that the fisherman should be constantly in contact with his fly. This is fairly easy for those who use traditional or attractor patterns; they will almost certainly be retrieving them at a reasonable pace and will know instantly if a fish takes hold. It is much more difficult for the imitative fisherman who either retrieves his artificial very slowly or even allows it to hang motionless in the water. Indeed, imitative patterns are very often intercepted by trout while sinking just after having been cast and such takes can be almost imperceptible. For these reasons, the imitative fisherman must rely far more on his eyesight than on his sense of feel, and should ensure that his tackle is so arranged as to make takes as visible as possible. If an unweighted artificial is fished on an ungreased leader, it may eventually sink to the required fishing depth but, while it does so, the leader and quite possibly the first foot or so of the fly line may well have sunk with it. If a fish takes such an artificial as it sinks through the water it is unlikely that the angler will notice it. If, on the other hand, a fairly heavily weighted artificial is used and the top half or two-thirds of the leader is lightly greased, the pattern

ALDER LARVA

CAENIS

Pheasant Tail Nymph	Gold Ribbed Hare's Ear
Last Hope	Grey Duster

CLARET DUN

Nymph   Pheasant Tail
    Spinner

CORIXA

CRANE FLY

DAMSELFLY NYMPHS
Early season

Mature

FRY

Polystickle
White Lure

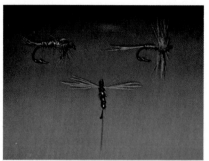

LAKE OLIVE
GHRE nymph   Dun
Spinner

FRESHWATER LOUSE

MAYFLY

Nymph   Adult

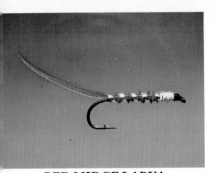

RED MIDGE LARVA

MIDGE PUPAE

Black   Red

Green   Brown   Phantom pupa

ADULT MIDGE (Blue Upright)

POND OLIVE

Nymph   GRHE

Greenwell's Glory   Spinner

SEDGE PUPAE

Yellow

Brown   Invicta

ADULT SEDGES

Richard Walker's

G and H sedge

SEPIA DUN

Nymph

Winged adult

SHRIMPS

Grey

Red

SNAILS

Black and Peacock Spider

Floating Snail

will sink to the required depth but the leader will now be drawn down progressively under the weight of the fly. Takes should be more apparent, the leader's descent either halting or accelerating, or the nylon twitching to one side.

It would be difficult to over-emphasise the slowness with which most imitative patterns should be retrieved. The only way in which the fisherman can really come to understand this is by watching the behaviour of the natural creatures that his artificials are designed to represent. The speeds concerned are deceptive. The smaller the insect or fish, the faster it will seem to move; but time it over a distance of (say) a foot and its sluggishness will immediately become apparent. Of course, speed of retrieve will often have to be something of a compromise. The Alder larva must be heavily weighted in order to reach the bottom in a reasonable time. If it is then fished at a realistically slow speed, it will snag the lake bed or catch in the weed; so it must be retrieved as slowly as possible but just fast enough to keep clear of the bottom. Flies dressed to swim upside down or with nylon 'brush off' filaments are less susceptible to these problems.

Where flies are to be fished close to the surface (rather than on it), they may either be held up by greasing the leader or by steady retrieval — the former only if the pattern is to be left static or retrieved very, very slowly. A greased leader moving across the surface at any speed creates line wake and is likely to disturb the fish. Knots in line or leader exaggerate this disturbance considerably. If the leader is meant to sink (as it should be if the fly is to be drawn along beneath the surface), then it should be rubbed down with a putty-like composition of Fuller's Earth, washing-up liquid and glycerine so that it cuts cleanly through the surface film.

The visible signs of trout taking an artificial nymph are

17

sometimes very clear — on occasion the whole leader and line may pull away from the fisherman. Often, however, they are difficult to detect and it is as well to strike at the slightest abnormality in the leader's behaviour without stopping to wonder whether it has been caused by a fish or not. The strike may result in both line and leader lifting back towards the angler without hindrance, but it is surprising how often it will be met by the solid resistance of a hooked trout.

Finally, those who have previously used a powerful rod and lures retrieved at speed will probably have become accustomed to the fish virtually hooking themselves. The light tackle, slow retrieves and gentle takes associated with imitative fishing require the angler to set the hook quite firmly. This is not done with any violence but the action should be a positive one and frequent sharpening of the hook point will pay dividends.

The Alder (*Sialis lutaria*)

Although most fly-fishermen are familiar with the Alder fly, few seem to know much about its larva. This is a little odd because, while fish take the larvae freely, the winged adults are of little consequence, particularly on stillwaters. The same goes for artificial patterns dressed to represent the respective stages in the insect's life.

Life history

When the Alder fly's eggs hatch amongst the vegetation at the waterside, the emergent larvae crawl or drop into the water where they burrow into the mud, appearing during the hours of darkness to feed voraciously on other aquatic insects. They swim badly but are capable crawlers and, when they are eight or nine months old they start to become more active and will sally forth in search of food

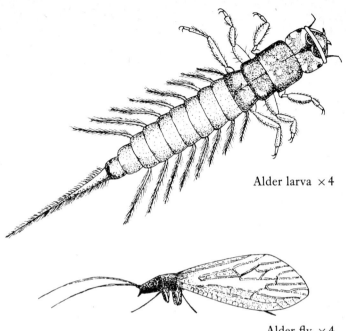

Alder larva × 4

Alder fly × 4

in daylight. A month or so later they should be fully grown. Between the beginning of April and late May or early June they crawl ashore to pupate and eventually hatch into adult flies. It is during this shoreward migration and during its earlier daylight feeding that the Alder larva provides an attractive mouthful for any trout cruising near the bottom.

The Alder larva's ferocious nature is matched by its ugliness. When fully grown the creature is generally about 1 in. long or very slightly less. Its buff head and thorax make up roughly a third of its overall length, the remainder consisting of a dull, steeply tapered, chestnut brown abdomen and a single long, spiky tail. It has six yellowish

19

legs which all emanate from the thorax and seven long, pointed tracheal gills protrude from each side of the abdomen.

Artificial pattern

The dressing that I have chosen to represent the Alder larva was designed by Stuart Crooks, an imaginative and inventive fly-tyer and a practical fisherman. It has provided us with some dramatic fishing from the beginning of the season until early June. This version is a slight modification of the original:

Hook	10–12 long shank
Tying silk	Buff or fawn
Underbody	Fine lead wire flattened horizontally with pliers
Tail	Light honey cock's hackle point
Abdomen	Chestnut seal's fur, ribbed with fine gold wire, trimmed short above and below (but not at the sides) and re-flattened
Thorax	Buff seal's fur (untrimmed) with buff raffia or raffine over the top

Presentation

The artificial sinks rapidly and should be fished on a floating line with a long (say 14 ft) leader, and should be worked very slowly along the bottom. On small lakes it should be fished in the deepest water, particularly early in the season. Trout frequently take this artificial as it sinks down through the water, and such takes will more readily be seen if the leader is lightly greased so that it sinks progressively under the weight of the fly. The pattern can be useful from the beginning of the season until early or mid-June and again in September or October when the naturals hatched earlier in the year have almost reached their full size. Late in the season, the pattern is most effective when fished early in the morning or just before nightfall.

The adult fly

The adult Alder fly is in evidence from May until mid-June. Although it is not unlike a mature sedge fly in general appearance it is, in fact, a member of a completely separate order, the Megaloptera, and unlike the sedges, has no hair on its wings. The fly is about $\frac{3}{4}$ in. long. Its head and legs are dark, almost black, and its shiny wings, which lie over its back like a roof when it is at rest, are mottled mid and dark brown in colour. While an artificial dressed to resemble the Alder fly can be effective on rivers, it is of no great value to the stillwater fisherman, probably because the naturals rarely fall onto the water in large numbers.

Caenis — the Angler's Curse

Of the five species of caenis or broadwing to be found in Britain only two, the Dusky and Yellow Broadwings, need concern the stillwater angler. Two of the remaining three are nocturnal while the third is exclusively a stream and river dweller. Not without reason is the caenis known to fishermen as 'the Angler's Curse'. Trout feeding on these insects late on a summer's evening are often very selective, and the situation can be exacerbated by the fish becoming preoccupied with the hatching nymphs or with one or other of the two species of adult when both are on the water together.

Life history

Caenis are members of the order Ephemeroptera — the upwinged or day-flies. Their nymphs are crawlers and live amongst the pebbles and stones at the bottom of the lake.

21

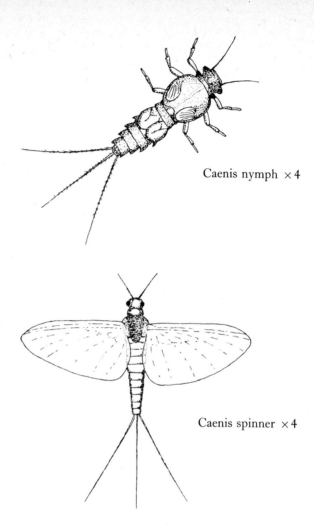

Caenis nymph × 4

Caenis spinner × 4

They rise to the surface during the early evenings between early June and August, the duns hatching out towards dusk. The transition from nymph to dun is rapid and, unlike many other hatching insects, the adults rarely seem to become trapped in the surface film for any great length of time.

The nymphs of both species are similar in appearance. They are $\frac{1}{4}$ in. long or slightly less, with rather tubby, mid-brown abdomens. Their dark brown thoraces are almost completely covered by even darker brown, almost black, gill plates and each nymph has three tails and six legs.

When it hatches, the Yellow Broadwing dun is just under $\frac{1}{4}$ in. long. Its single pair of wings, a very pale, translucent grey, are rather wide for their height (hence the insect's colloquial name). The thorax upon which the wings are mounted is almost black on top and a lighter, rather dirty brown colour underneath. The abdomen is a pale, pastel yellow and the insect has three tails.

Less common than the Yellow variety, the Dusky Broadwing is a little larger than its counterpart and looks very much like it except that its abdomen is a creamy grey colour.

The spinners of both species are, for all practical purposes, similar in appearance to the duns.

Artificial patterns

The caenis nymph can be effectively represented with a small Sawyer's style Pheasant Tail Nymph. The dressing is as follows:

Hook	16 down-eyed
Tying 'silk'	Very fine copper wire
Tail	The points of three cock pheasant centre tail feather fibres, about $\frac{1}{8}$ in. long
Abdomen	Cock pheasant centre tail feather fibres wound into a rope with fine copper wire

| Thorax | Built up from fine copper wire |
| Wing cases | The dark butts from the cock pheasant fibres doubled over the thorax |

For the hatching nymph I generally use a very small Gold Ribbed Hare's Ear dressed as follows:

Hook	16 fine wire, up-eyed
Tying silk	Pale primrose
Whisks	A small bunch of hare's fur guard hairs
Body	Dark fur from the root of a hare's ear ribbed with fine gold oval tinsel
Hackle	A few long strands of the body material picked out with a dubbing needle

Of the various patterns available to suggest the adult càenis duns and spinners, two dressings, the first by John Goddard, the second the ubiquitous Grey Duster, have provided me with some success during rises to Broad-wings — although it should be said that success is a relative word where this infuriating insect is concerned. The dressings are:

Last Hope (John Goddard):

Hook	17 or 18 fine wire, up-eyed
Tying silk	Pale yellow
Whisks	6–8 fibres from a honey dun cock's hackle
Body	Two or three Norwegian goose or condor herls — grey buff
Hackle	Dark honey; very short in the fibre

Grey Duster:

Hook	16 fine wire, up-eyed
Tying silk	Purple
Whisks	A small bunch of badger cock's hackle fibres — or none

Body	The blue under fur from a rabbit
Hackle	Badger cock's — very short in the fibre

Presentation

The nymph should be fished very slowly, close to the bottom, particularly over a clean, stony bed, or in the vicinity of weed, during the early evenings from June to August when a hatch of caenis may be expected later. Takes are often very gentle indeed and the top two or three ft of the leader may be lightly greased to make spotting them a little easier. Although the nymph is tied with copper wire, it is so small that it sinks very slowly and the whole of the leader apart from the greased butt length will have to be cleaned with a mixture of Fuller's Earth and washing-up liquid if the fly is to go down deep enough.

Trout taking Broadwing duns and spinners frequently do so in a series of head and tail rises. If a hatch is apparent, the angler should fish one of the artificials given above on a greased leader, casting it into the paths of rising fish. Should this fail repeatedly it may be that the trout are taking the hatching nymph; then the small Gold Ribbed Hare's Ear fished in, rather than on, the surface film may do the trick.

The Claret dun (*Leptophlebia vespertina*)

How this insect came by its colloquial name is something of a mystery for while the spinner of the species sometimes has a noticeable claret tinge to its body, the dun is a very deep chestnut brown colour. The Claret dun, a creature of high summer, is widely distributed throughout Britain and Ireland but shows a general preference for peaty, acidic waters. Both C. F. Walker and John Goddard have noted that it is most commonly to be found in Ireland

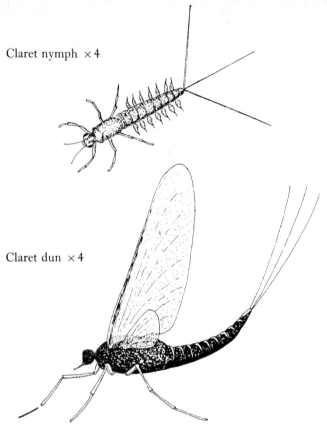

Claret nymph × 4

Claret dun × 4

but as more stillwaters have become available in the West
Country, it seems to have increased in numbers there too.

Life history

The Claret nymph is distinctive in appearance and, once
identified, could only ever be confused with its close
cousin, the Sepia nymph. Its body, about ½ in. long, is
fairly broad by comparison with the nymphs of most other

upwinged flies and is a very dark reddish-brown colour. The insect's abdomen is fringed on either side by seven pairs of long, pointed gill filaments and its three tails, each half as long again as its body, are widely spread. As they develop, the wing cases change from dark brown to black.

The Claret nymph is a poor swimmer and spends most of its life crawling about amongst the weed and decaying vegetation on the lake bed. Unlike the Sepia nymph, it lives in shallow water and J. R. Harris has pointed out that in lakes where the bed alters in composition from area to area, it is mainly to be found where the bottom has a peaty make-up.

The insect's season is a somewhat open ended one but the main hatches occur between mid-May and mid-July, normally around midday and in the early afternoon. Once the nymph is fully mature, a layer of gas builds up beneath its skin, carrying it to the surface and initiating the splitting of its case which allows the dun to emerge. This is a common occurrence amongst upwinged flies but is more apparent in slow moving, bottom dwelling nymphs than in the more accomplished swimmers.

The body of the Claret dun itself, just over $\frac{1}{4}$ in. long, is a very dark reddish-brown colour. The insect has four wings and, as is usual in upwinged flies, the front ones are very much larger than the back ones. In this case, the fore-wings are a dark, smoky grey and, in contrast, the rear ones are a much lighter, pale fawn colour. The fly has three tails, each half as long again as its body.

The spinner, most commonly seen in the evening, is similar in size to the dun. Its body is glossy and very dark indeed, almost black, with a slight deep red tinge, particularly towards the tail. Its abdomen is ringed in a lighter shade at the segmentation joints and its wings, bright and transparent, are faintly veined in a pale shade of watery brown.

Artificial patterns

Of the various patterns designed to represent the Claret nymph, those which incorporate a silver rib seem to be the most successful. This may be because the tinsel produces an effect not unlike that imparted by the gaseous layer beneath the real nymph's skin as it prepares to hatch. The nymph pattern given in the section on the Sepia dun later in the book may also be used to represent the Claret nymph but should be dressed on a slightly smaller hook, say a no. 14, in this case.

Although various dressings have been invented specifically to represent the Claret dun, my own experience suggests that a standard dry Pheasant Tail is as effective as anything, particularly if the fibres used for the body are fairly dark. The dressing that I use is as follows:

Hook	14 fine wire, up-eyed
Tying silk	Brown
Whisks	A small bunch of rusty dun cock's hackle fibres
Body	Three cock pheasant centre tail feather fibres (the darker the better) ribbed with fine gold wire
Hackle	Dark blue dun cock's

(G. E. M. Skues's dressing, which is more suitable for lighter brown upwinged flies, is given later in the section on the Sepia dun.)

Commander C. F. Walker, J. R. Harris, John Henderson and others have all designed patterns to represent the Claret spinner. While I am sure that they are efficient, each employs dark claret or dark brown seal's fur for the body. In the natural, both thorax and abdomen are almost black and their outlines are hard and sharp. For this reason, I prefer to use a copy of a dressing of the spent natural which I bought in Ireland some years ago and which has served me well. I do not know who originated it:

Hook	14 fine wire, up-eyed
Tying silk	Black
Tails	A reasonably large bunch of honey dun cock's hackle fibres
Body	Dark claret floss silk ribbed with fine gold wire
Wings	Bright, light honey dun cock's hackle, wound fairly full, drawn into two equal bunches, one on either side of the shoulder, and held in position with a figure of eight binding
Hackle	None

Presentation

The nymph can profitably be fished from the middle of May onwards. When no duns are hatching it should be worked very slowly along the bottom on a floating line, with only the top foot or so of the leader greased, in shallow water. Because of the insect's predeliction for acid environments there may be some advantage in concentrating on areas where the banks are peaty, or on the still water around the inflows of moorland streams. However, although it is true that the naturals prefer these areas, I would hesitate to suggest that use of the artificial should be confined to them. During a thin hatch of duns, when trout are not feeding on the surface, the nymph may produce results if fished with a slow 'sink and draw' retrieve.

During a real hatch of Claret duns the trout can become quite preoccupied with the newly emerged adults which sit on the surface drying their wings for an appreciable time before flying off. At such times the artificial, fished on a floating line and a greased leader, cast to individual fish and left motionless on the surface can produce outstanding sport.

The same technique is used for the spinner as for the dun. Falls of spinners generally occur from early evening onwards and the artificial should be allowed to float in the surface film rather than on it, in the form of the spent insect.

Corixids

While autopsies suggest that only a relatively small percentage of trout have a taste for corixae (or Lesser Water Boatmen), these insects are significant to the fisherman because they are present throughout the season and because they are available in considerable numbers during the mid to late summer when many other insects are in fairly short supply.

The naturals

There are more than 30 species of corixa and while some are too small to be of much interest to the angler, others grow to $\frac{1}{2}$ in. or more in length. The adults lay their eggs on underwater plants between January and March and the nymphs hatch three weeks to a month later, depending on the temperature and the species. The corixid nymph is very similar in appearance to its adult counterpart but its wing cases are translucent, allowing the body markings to show through, and there may be some insignificant variations in shape and colour. It reaches maturity in three to four months.

The adult is beetle-like. Its body, egg shaped in plan view, is markedly flattened horizontally — a fact rarely taken into account by fly-dressers. While the front two pairs of legs are inconspicuous, trailing beneath the insect as it swims, the third are sturdy, paddle-like appendages

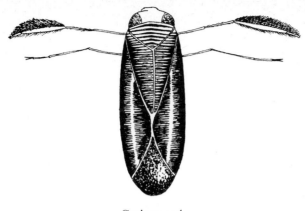

Corixae × 4

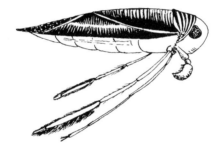

which the corixa uses to power itself through the water. The wing cases covering the insect's back may vary in colour from dull yellow ochre through olive to dark brown, and are often mottled or geometrically patterned. The corixa's body, a pale creamy white, is often almost completely obscured by the silvery air bubble which surrounds it.

Corixae generally live close to the bottom and are commonly found in and around clumps of weed. They rarely venture into water more than five or six ft deep. In simple fly-dressing terms the insect is relatively easy to portray, but its movement is an altogether different matter. Having no gills, corixae rely on air bubbles collected from the surface for their oxygen supplies and have to make regular journeys to replenish these bubbles. They swim upwards with a remarkably rapid, jerky, zig-zag movement. Because their renewed air supplies make them buoyant, the return journey is a noticeably more laborious affair although they still tend to follow a very erratic path. Even when simply pottering around, corixae swim very busily in short, sharp jerks. They are at their most vulnerable when away from the cover of the weed in which they live, particularly as they struggle down in their silvery, gaseous cocoons, and it may be that their peculiar darting movement is an evolved defence against predators; it certainly does nothing to assist the fisherman whose artificial can only sink vertically, or almost so, and who must retrieve his pattern in an almost straight line.

In the mid to late summer corixae actually take to the wing and are able to fly quite considerable distances over the water at surprisingly high speeds.

Artificial pattern

There are several effective dressings designed to represent the corixa. The one given below seems to me to be particularly realistic and has caught many fish:

Hook	10 or 12, down-eyed
Tying silk	Pale primrose
Tag and rib	Medium silver tinsel
Wing cases	Speckled brown hen's quill slip tied over the completed body and given at least four coats of fine polyurethane varnish
Underbody	Fine lead wire, flattened horizontally
Body	White floss silk ribbed with medium silver tinsel

Forelegs	One turn of pale brown hen's hackle tied beard fashion
'Paddles'	Two cock pheasant centre tail feather fibres tied to stand out at right angles from the shoulder

Presentation

This pattern, which should be fished in the smaller size at the beginning of the season and the larger one from June onwards, sinks quickly. As the fly is very often taken 'on the drop', the top two-thirds of the leader should be greased so that it sinks progressively under the weight of the fly. The artificial may be cast either to individual fish or to likely lies around weed beds. Once it has reached the required depth, it should be retrieved in sharp one or two in. pulls with a pause every 12 in. or so to allow it to sink again.

Crane flies (*Tipulidae*)

Crane flies, or Daddy-long-legs as they are more generally known, are a familiar sight both on and off the water during the late summer. There are, in fact, several hundred species of 'Daddy', some of them less than $\frac{1}{4}$ in. long, others more than an inch from head to tail. They range in colour from dusky grey through every shade of brown to yellow; some are mottled or patterned. It would obviously be just as impractical to describe each species that could be of interest to the stillwater flyfisherman as it would be unnecessary to dress patterns to match them. However, the Crane fly arrives at a most propitious time of year and the angler who has an artificial or two in his fly box may well avoid blanks on some of those hot, still, August afternoons when the fish seem generally lethargic and uninterested.

33

Life history

Although most species of Crane fly are terrestrial, some spend the pupal phases of their lives in moist ground near the water's edge and a few are actually aquatic, spending

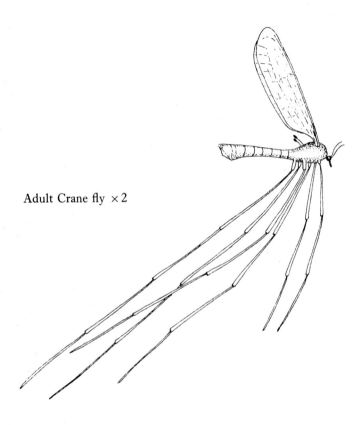

Adult Crane fly × 2

both their larval and pupal stages in the water. In general terms, the adults of the commonest Crane flies to be found around stillwaters are about ¾ in. long in the body. Their abdomens are flat yellow ochre or grey-brown in colour while their thoraces are normally slightly darker. Their single pairs of wings, translucent and often with a faint red-brown or greyish tinge, approximately equal their abdomens in length and, needless to say, their legs always look disproportionately long and gangling.

Crane flies are remarkably poor fliers and, once airborne, are largely at the mercy of the wind. They trail their legs as they fly and when they eventually force-land on the water, they lie in the surface film with their legs spread out behind them.

Artificial pattern

Several patterns have been designed to represent the adult Crane fly. The one given below should serve most purposes and has the advantage of being relatively simple to tie:

Hook	12 or 14 long shank
Tying silk	Brown
Body	Four cock pheasant centre tail feather fibres ribbed with fine gold wire
Legs	Six to eight cock pheasant centre tail feather fibres, knotted to represent leg joints, tied in by their butts at the shoulder with points trailing behind the body
Wings	Iron Blue dun cock's hackle points tied 'spent'
Hackle	Two stiff, bright, red-brown cock's hackles

Presentation

The artificial may be fished in any one of three ways when the naturals are about between mid-July and mid-September. Warm afternoons seem to be most productive, particularly if they follow showers or light rain.

Perhaps the most traditional technique for fishing the Daddy-long-legs is dapping. For most people dapping is associated with the Irish loughs, but there is no good reason why it should not be just as effective in this country. A longish rod and a floss or nylon monofilament line are used in a reasonable breeze. The line is allowed to blow out over the water and considerable expertise is needed to keep the fly just tripping over the surface. Trout take the dapped 'Daddy' with a gentle suck, and some self discipline is required to delay the strike for long enough for the fish to take the fly fully into their mouths.

The artificial Crane fly may also be fished as an ordinary dry fly on a greased leader. The angler should fish with the wind on his back. While the artificial will sometimes be taken while lying static on the water, the trout are attracted to it more often if it is retrieved slowly in short twitches. As with the dap, the trout generally sip the almost static artificial down and the strike must be delayed. Obviously, if large numbers of natural Crane flies are being blown onto the water and the fish are feeding on them seriously, it is possible to cast to individual rising fish. Otherwise, fishing the water can sometimes be quite productive.

Finally, an artificial Crane fly will often be taken with a wallop if it is retrieved in quick, 12 in. pulls so that it creates a bow wave on the surface. I am not sure, however, that the fish are not taking the fly for a sedge under these circumstances.

Damselflies (*Zygoptera*)

Damselflies, with their long, slender, brilliantly coloured bodies, are a familiar sight in summer, particularly on smaller stillwaters with plenty of marginal vegetation. They are similar to dragonflies in appearance but, unlike dragonflies, fold their wings over their backs when at rest. The commonest have bright blue bodies, others wear red, green or golden olive livery. While trout may sometimes be seen slashing at the egg-laying females, these insects are rarely found in autopsies. For this reason, and because the adult damselfly is too big to be represented on any reasonable size of hook, the winged creature is of interest only to the most experimental and innovative fisherman. However, the damselfly nymph is an altogether different matter and is taken freely by trout. Although it is available throughout the first four or five months of the season, it spends much of its time impersonating the weed on which it perches and does not become readily accessible to the fish until mid-summer. Nevertheless, an artificial damselfly nymph can be a useful standby from the beginning of the season onwards and its usefulness increases until the adults finish hatching in mid-August.

Life history

While the fully grown nymphs range in colour from dull brownish yellow through various shades of brown to deep green according to species, they all seem to be a fairly uniform pale, translucent weed-green colour when young. In March or April they are generally about ¾ in. long. Their thoraces constitute about a quarter of their overall body lengths and their abdomens are only slightly tapered. Each nymph has three pointed, translucent, leaf-like 'tails' which are in fact tracheal gills, and six sturdy legs emanating from its abdomen.

By mid-summer the damselfly nymph may be as much as an inch and a half long and proportionately more solid than it was as a youngster. Shortly before the adult hatches, a very apparent brown case forms over the top of the thorax and wing pads, angled backwards over the abdomen, sprout from the back of the thorax.

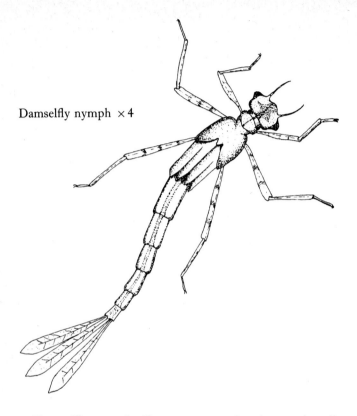

Damselfly nymph × 4

Damselfly nymphs live among weed and are primarily crawlers, although they can swim and sometimes do so using a curious, snake-like movement of their abdomens to propel themselves through the water. They are highly carnivorous and use their powerful, extendible jaws to catch other insects which share their weedy habitat. When the time comes for the adult to hatch, the nymph either climbs up vegetation which protrudes above the surface or swims ashore. Once clear of the water, it splits along its back and the winged adult emerges.

Artificial patterns

Because of the pronounced differences between damsel-fly nymphs early in the season and later in their lives, it is worth having two patterns with which to represent them. The first one given below should serve from March to the beginning of May on any water where the naturals are known to exist:

Hook	8 down-eyed
Tying silk	White
Gills ('tails')	Three pale olive cock's hackle tips
Body	Pale green seal's fur, dressed very sparsely and trimmed to taper slightly to the tail — ribbed with fine silver wire
Legs	One turn of olive hen's hackle

The second pattern, to represent the adult nymph, should, where possible, be dressed to match the colour of the predominant naturals in the water. If in doubt, olive and dark green are the most likely shades:

Hook	8–10 long shank
Tying silk	Green
Tail	Three medium olive cock's hackle points about ¼ in. long
Abdomen	Olive, green or brown seal's fur ribbed with fine gold oval tinsel
Thorax	As abdomen, unribbed, with eight strands of cock pheasant centre tail fibre over the top
Wing cases	The butts of the pheasant tail fibres turned back and divided so that they project for about ⅛ in. above either side of the abdomen
Legs	The points of the eight pheasant tail

	fibres divided, four to each side
Head	Built up with tying silk and well varnished

Presentation

The pattern representing the young nymph is particularly effective when cast to individual trout. Alternatively, it may be fished very slowly, close to the bottom and around the edges of weed beds. The same applies to the more mature pattern which, in addition, can be effective if fished an inch or so beneath the surface with a steady retrieve when the adults are in evidence.

Fry

Many fishermen are surprised to discover how ferocious trout can be, particularly towards the end of the season when they are building up their strength for winter and for spawning. Coarse fish fry — roach, perch and bream — as well as sticklebacks tend to congregate in shoals and provide easy pickings for marauding trout during the late summer and early autumn. At this time of year an imitative artificial may not only provide dramatic fishing, it may also tempt some of the larger fish which would be difficult or even impossible to attract with a nymph or a dry fly.

There are those who would say that to use a fly representing a small fish is little better than spinning, but this is an unsupportable assertion. There is very little that is realistic about a spinner which throbs through the water in an effectively straight line and which, when taken, has almost certainly aroused some curious or aggressive instinct in the fish, rather than the feeding one. An artificial minnow, built from light, fly-dressing materials and fished with a fly rod and line, can be made to resemble

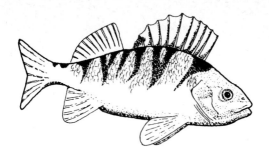

Perch and bream fry × 1

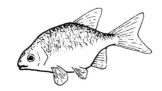

Stickleback × 1

closely the natural creatures upon which the trout are feeding — and this is what fly fishing is all about.

The naturals

All our trout reservoirs have their own populations of coarse fish, particularly perch, roach and bream. These fish spawn in the spring. The young grow quickly during the first year of their lives and may be as much as two inches long by August or September. The fry live on the nymphs and larvae of aquatic insects and are therefore to be found where these creatures abound — in vegetated shallow water, around weed beds and near the vertical surfaces of piers, jetties and structures such as the valve towers in reservoirs. Sticklebacks, which grow to a maximum length of about three inches, are essentially shallow water fish and are of particular interest to fishermen at reservoirs like Chew where the margins are gently shelving in many places.

All these small fish swim in shoals. The trout attack them by charging through their ranks, taking in any stragglers left behind when the silvery mass parts in panic, and then returning at a more leisurely pace to collect any fry that may have been crippled or injured in the general pandemonium.

It is important to realise that sticklebacks and fry do not swim quickly. To the human eye, they may appear to do so but this is an illusion caused by their diminutive size. When undisturbed, they idle along at a very sedate pace. When alarmed, they swim in short darts, pausing briefly every foot or so.

Artificial patterns

Many patterns have been specifically designed to represent fry, and many more lures and attractor patterns are believed by their users to be accepted by trout as small fish. The Polystickle, designed by Richard Walker, is both realistic and effective. Indeed, so killing is it under the right circumstances that it has become a standard part of the stillwater angler's armoury. It is particularly useful when trout are feeding on sticklebacks or small perch fry:

    Hook   8 long shank (silvered)

Tying silk	Dark brown or olive
Underbody	Front third, red floss silk; rear two thirds, brown or olive floss silk ribbed with broad silver tinsel
Body covering	Stretched PVC strip built up into a fish shape
Back and tail	Darkish raffine moistened and tied in before the body is built, brought down over the body and tied in at the head; cut and shaped at the tail
Throat	Scarlet cock's hackle fibres tied in as a false hackle
Head	Built up with tying silk and well varnished

This second fry-representing pattern, the White Lure, has been chosen because of its effectiveness when trout are attacking shoals of small roach or bream:

Hook	6–12, two or more tied in tandem
Tying silk	White or olive
Bodies	White floss ribbed with oval silver tinsel
'Wings'	White cock's hackles

Presentation

When trout may be expected to be feeding on fry, it is important to locate the shoals of small fish. Fortunately, this is often fairly easy. As the trout harry them, the fry panic and are frequently to be seen leaping clear of the water, giving the appearance of a handful of gravel thrown onto the surface. If you can approach close enough, confirmation of the cause of the commotion will sometimes be available in the form of floating dead or dying fish. Shoals further away can quite often be pinpointed by wheeling and stooping flocks of sea birds which feed on the crippled fish, particularly on the larger reservoirs.

Once a shoal of fry upon which trout are feeding has been located, the artificial can be fished in one of two ways. It can be cast beyond the shoal and retrieved very slowly through it. As a trout charges in on its attacking run the shoal will scatter, leaving the artificial for the trout. Takes when fishing like this can be very violent and the leader must be sufficiently strong to withstand a much heavier impact than usual. Alternatively, if a trout has been seen harrying the shoal, the artificial may be fished very slowly indeed, with a series of minute twitches to simulate the movement of an injured fish. This can be a deadly technique on occasions and the takes are generally relatively gentle. For both these methods, the artificial should be fished on a floating line and an ungreased leader.

If no shoals of fry are in evidence, it is often worth fishing an artificial slowly along the edges of steep-sided weed beds.

For trout that may be feeding on the shoals of fry around piers, valve towers and so on, it may be necessary to use a sinking line. Even so, the retrieve should be calculated and slow. I am convinced that although fry representing patterns can take fish when stripped through the water, they are no more than 'attractor patterns' when used in this way.

The Lake Olive (*Clöeon simile*)

The Lake Olive is common on stillwaters throughout the British Isles but shows a marked preference for alkaline rather than acid or peaty ones. As nymphs, duns and spinners, these upwinged flies constitute a significant proportion of many trouts' diets and it is strange that relatively little importance seems to be placed on them by the modern generation of stillwater fishermen. Lake Olives are in evidence throughout the year although they are most plentiful at the beginning and end of the summer. In cool weather they generally hatch around midday or in the early afternoon, but on hot days they may postpone their appearance until the evening.

Life history

The Lake Olive nymph is about ½ in. long including its tails. Its abdomen and legs are mottled yellow and brown while the substantial gill plates covering its thorax are almost black. The nymph lives amongst patches of moss and short-stemmed weed but can swim very fast for short distances when in the open water. It is most vulnerable to predation by trout when it becomes buoyant and floats to the surface to hatch into a dun.

Lake Olive duns which hatch in the early summer are markedly larger and darker than those that hatch later in the season. The earlier dun, just under ½ in. long in the body, has a dark grey-green thorax, a medium or dark olive-green abdomen and two grey tails. Its wings are a dusky grey, tinged with green at their bases. The later arrivals have paler, rusty-olive bodies and their wings have a more gingery appearance.

As is the case with all upwinged flies, the Lake Olive spinners are very much brighter than the duns. Their thoraces are very dark brown, tinted with mauve, and their abdomens are a dark red-brown on top and greyish green underneath. The spinners' wings are glossy and translucent and are coloured with a very pale brownish tinge.

Although the Lake Olive is as commonly to be found on large lakes as on smaller ones, it generally prefers to live in sheltered areas, bays and inlets, rather than along exposed or rocky shorelines. It seems to prefer water of medium depth — between 6–14 ft — and a bed where thick moss or dense short-stemmed weeds abound.

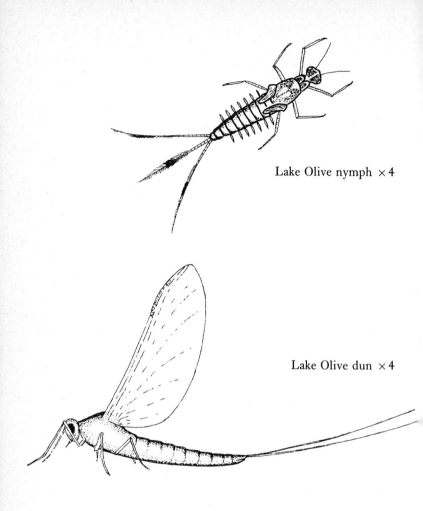

Lake Olive nymph × 4

Lake Olive dun × 4

Artificial patterns

A most useful pattern to represent the Lake Olive nymph is the American Gold Ribbed Hare's Ear Nymph as dressed by David Collyer (and in which no hare's ear is actually used!):

Hook	12 or 14 down-eyed
Tying silk	Brown or black
Tail	Hare's body hair
Body	Hare's body hair ribbed with oval gold tinsel
Thorax	Hare's body hair — a few strands picked out to represent legs
Wing cases	Dyed black turkey tail

Most of the recognised patterns to represent the Lake Olive dun incorporate quill fibre slip wings which, while helpful in cocking the fly and possibly making it more visible, seem to me to be unrealistic from the fish's viewpoint. My own hackled dressing is as follows:

Hook	14 fine wire, up-eyed
Tying silk	Brown
Whisks	A small bunch of light dun cock's hackle fibres
Body	Dark olive condor herl ribbed with silver wire
Hackle	Two medium blue dun cock's hackles

There can be few better patterns to suggest the Lake Olive spinner than that given by C. F. Walker in *Lake Flies and Their Imitation*. His dressing is:

Hook	12–14, fine wire, up-eyed
Tying silk	Black
Whisks	Fibres from medium blue dun cock's spade or saddle hackle
Body	Dark red seal's fur ribbed with gold tinsel
Wings	Pale, brassy dun cock's hackle points, tied 'spent'
Hackle	Pale brown or honey dun cock's hackle — or none

Presentation

The Lake Olive nymph may be used throughout the season. In the absence of hatching duns to show the whereabouts of the naturals, it should be fished fairly deep, amongst the weed in sheltered bays. As the natural moves in short darts when away from weed cover, the artificial should be retrieved in sharp two or three inch pulls with pauses between them. The margins should not be ignored and, indeed, should be fished out before the deeper water is tried, particularly when duns are hatching.

The dun and spinner should be cast to rising fish, or into areas in which fish are moving, and allowed to float undisturbed. They should be treated with a floatant so that they sit high on the water. Both dun and spinner seem to become markedly less attractive to the fish when waterlogged.

Lice (*Assellus*)

The freshwater louse is another creature that seems to have been largely ignored by modern stillwater fishermen. This may be because its bottom dwelling, slow moving habits are difficult to reproduce with an artificial pattern on the end of a line. Nevertheless, so prolific and widespread is the louse, and so frequently do trout feed on it, that a little thought and effort devoted to representing it can be dramatically repaid.

The naturals

There are several species of aquatic louse, but for all practical purposes the fly fisherman may quite safely treat them as one. The adult insects, which may grow to $\frac{1}{2}$ in. or more in length, are not dissimilar in appearance to their terrestrial cousins the wood lice. Their bodies, oval in plan

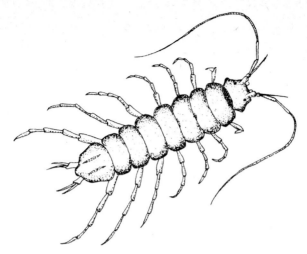

Freshwater louse × 4

view and noticeably flattened horizontally, range in colour
from a dirty white to dark brown or dusky grey and are
clearly segmented. Each insect has seven pairs of lateral
legs, the three rearmost ones being markedly longer than
the front four, and an additional pair of long, forked ones
at the rear. The two slender antennae emanating from the
louse's head are about half as long as the body.

Freshwater lice have no intermediate stages in their life
cycles and their young are simply miniature versions of
the adults. Nor do these animals seem to have a particular
breeding season, so lice in a variety of stages of growth
may be found together in the water at any one time and
are available to trout throughout the year.

Perhaps surprisingly, freshwater lice are non-swimmers.
Pottering around in fairly shallow water, they spend their
lives amongst mud or gravel, in rotting or rotten leaves and
weed or crawling about rather ponderously on pads of

moss and the stems of aquatic vegetation. They appear to dislike running water and are therefore rarely found near inflows or outflows.

Artificial pattern
Very few patterns have been designed to represent the freshwater louse. Although Commander C. F. Walker gives a dressing in his *Lake Flies and Their Imitation*, I have never been able to find partridge hackles (which he advocates for use as legs) that are long enough in the stem and short enough in the flue to provide anything approaching realistic proportions. However, I have had some considerable success with Commander Walker's artificial tied over a flattened lead underbody and with the partridge hackle omitted altogether. Many of the trout thus caught have been found to contain large numbers of natural lice in subsequent autopsies. The dressing for this non-descript looking creation is as follows:

Hook	12 or 14 down-eyed
Tying silk	Fawn or brown
Underbody	Fine lead wire, flattened horizontally
Body	Mixed brown and grey hare's ear, trimmed short on top and ribbed with fine silver tinsel

Presentation
This louse pattern is a useful standby throughout the season when there are few indications as to what the trout may be feeding on. I find that I generally use it most during the first couple of months before the weed has grown; thereafter, it tends to get caught in the aquatic undergrowth. It should be fished on a floating line and an ungreased or very lightly greased leader, the latter being long enough to allow the fly to sink right to the bottom.

The retrieve should be very slow and steady and the artificial should be kept as close to the bottom as possible. The margins should always be fished before the deeper water is tried (a good general rule regardless of the pattern) and the cast should always be fished right out; I have watched trout follow this louse pattern for a considerable distance and then take it almost under the rod tip. It should always be remembered that when fishing close in, keeping out of sight will pay enormous dividends.

Mayflies (*Ephemera spp.*)

Mention the Mayfly to any English fisherman and the chances are that his thoughts will turn to southern chalk streams. Although some of the finest fishing in these islands is to be had on the great Irish loughs during the Mayfly season, and several of our own alkaline lowland lakes and reservoirs produce respectable hatches of their own, many people seem to have a curious inbuilt resistance to trying a Mayfly on stillwater.

The Mayfly is the largest, best known and visually most distinctive of all the upwinged or day-flies. Where hatches are sparse the trout often seem to be nervous of so outsized a winged adult. Under these circumstances, a careful imitation of the nymph can sometimes be very effective. Where the flies are prolific, the fish will take them with gay abandon once they have become accustomed to their presence.

Three species of Mayfly are to be found in the British Isles, two fairly common and the third markedly less so. Fortunately, both of the former are so similar in appearance that for fly dressing purposes they may be treated as one species. Whether the Mayfly has a one or two year life cycle matters very little to the practical fisherman; the pundits will doubtless go on debating the question for ever. What is beyond dispute is that the insect goes through all the four stages that are typical of the upwinged flies — egg, nymph, dun and spinner.

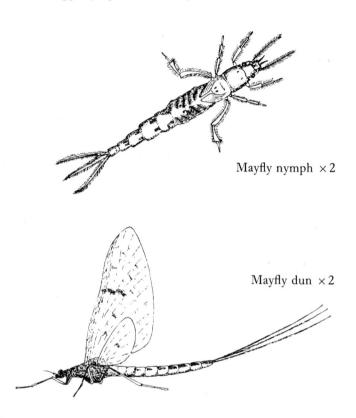

Mayfly nymph × 2

Mayfly dun × 2

Life history

The Mayfly nymph is about 1 in. long when fully grown and its body is normally a very pale fawn colour; where it lives against a darker background it may have a brownish tinge to it. It has six sturdy legs, three tails and, by the time it is of real interest to the fisherman, well developed, dull brown wing cases over its thorax. Prominent feathery gills project laterally from all but the last three segments of its abdomen and are held curved over its back. This unattractive creature is primarily a crawler and spends most of its life burrowing in the sand or gravel on the lake bed. A week or two before hatching the nymph leaves its shelter and, of course, becomes particularly vulnerable to predation by trout. When ready, it swims directly and fairly rapidly to the surface where the dun hatches.

There are visible differences between male and female Mayfly duns but they are relatively slight and need not concern us here. Entomologists will forgive me for describing a somewhat hermaphrodite creature. The dun is about $\frac{3}{4}$ in. long in the body and its three dark grey tails are about twice as long again. Its thorax and head are very dark brown, almost black, while its abdomen is a pale creamy grey colour with slate grey or brown markings on top. At rest, it supports itself on six necessarily sturdy brown legs. Its wings, one large pair and one small, are a rather dull, lustreless pale greyish-yellow colour, heavily veined in brown and with dark olive brown blotches on them.

The female spinner (for it is almost always the female that finds its way onto the water to become potential trout food) is much the same size as the dun but is a considerably more brilliant creature. Again, the head and thorax are very dark brown, the abdomen is creamy white with dark brown rings around the last three segments and faint markings on top of the others. Its legs and tails are dark greenish-brown. The spinner's wings are brilliantly transparent with heavy, rust coloured veining, and usually have a few small, dark spots on them.

Artificial patterns

Several patterns have been developed to represent the Mayfly nymph, those by Richard Walker and David Collyer being particularly well known. Whichever you choose, they are most killing artificials and often seem to tempt the larger fish. The dressing given below has the advantages of sinking quickly and being imbued with considerable inherent movement and 'life':

Hook	8–10 long shank
Tying silk	Brown
Tails	Cock pheasant tail fibre points
Underbody	Fine lead wire flattened horizontally
Abdomen	Rear half — buff condor herl; front half — dull yellow seal's fur. The whole ribbed with black nylon monocord.
Thorax	Dull yellow seal's fur
Wing cases	A slip of speckled brown hen's quill
Legs	One turn of brownish partridge hackle

There are so many patterns to represent the winged adult Mayfly that it is difficult to know which to select. After due consideration, I have chosen a hackled version which may be as effectively used for dapping as for more conventional fly fishing, and which is truly buoyant — essential when so large a hook must be used. I do not know who originated the dressing or where but it is as follows:

Hook	10 up-eyed
Tying silk	Black
Tail	Four or five cock pheasant tail feather fibres
Body	White raffine ribbed with black nylon monocord

Hackle	A medium blue dun cock's hackle between two badger cock's hackles, very full

Presentation

Mayflies hatch between the third week in May and the first in June, depending on geographical area and, perhaps, the severity of the preceding winter. Their season tends to be a short one and the only way of predicting it accurately on any particular water is to seek the advice of knowledgeable locals.

The nymph may realistically be fished from about ten days before the Mayflies themselves are due to appear until a day or so after the hatch actually starts. At the beginning of this period the nymph should be fished slowly, close to the bottom on a long leader and with a floating line. As the first duns appear, it should be allowed to sink and then retrieved in a series of steady pulls, fast enough to bring it up towards the surface.

Once the fish have lost their wariness of the winged adult naturals, the hackled Mayfly can be used. Perhaps one of the most rewarding techniques is dapping, either with one or more natural insects clipped to a specially designed hook, or with the artificial. With a long rod and a light, floss line, the fly is tripped across the waves as the boat drifts down the lake on the breeze; movement of the fly is all-important. When the fly is taken, it is essential to pause for two or three seconds before setting the hook.

Or the hackled pattern may be fished as a conventional floater, either from the bank or from a moored boat. Used thus, it may be given a little life with an occasional twitch of the line. Here again, the strike must be delayed for as long as it takes the fish to turn down with the fly in its mouth.

Midges (*Chironomids*)

Call them buzzers or chironomids if you will, the midges (or rather, their pupae) are by far the most important of all our stillwater trout flies. Spread throughout the British Isles, present in substantial numbers wherever they occur and with one species or another hatching at every stage of the season, they are bread and butter to the trout. And in this the fly fisherman is fortunate, for taking fish on an imitative artificial when they are feeding on the hatching insects is a fascinating and rewarding exercise.

Almost 400 species of midges have been identified in the United Kingdom and it would obviously be both impractical and unnecessary either to describe them all here or to dress artificials to match any great number of them. Anglers owe a debt of gratitude to John Goddard for the painstaking research that he carried out on the chironomids during the 1960s and those who wish to study the subject in detail would be well advised to turn to his *Trout Flies of Stillwater*. I am limiting myself to a description of the general characteristics of the insects at the three significant stages in their lives and to the provision of a relatively small selection of dressings that should meet most requirements. The dressings themselves can be related to the seasonal table at the end of the book.

Life history

The eggs laid by the female midges on the water's surface remain there until the larvae emerge and sink to the bottom three or four days later. These larvae, which may grow to be as much as 1 in. long, range in colour from an almost transparent pale grey through yellow ochre and brown to green and red, depending upon the species, with green and red being the commonest hues.

Midge larva × 4

Midge pupa × 4

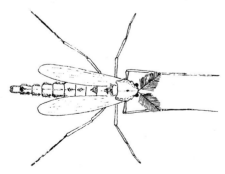

Adult midge × 4

57

The larvae are worm-like in appearance (hence the name 'blood worm' for the red ones) with clearly defined, almost bulbous, segmentation. While some build tubular cases of mud or sand for protection, others are free-swimming and propel themselves with a peculiar lashing motion, repeatedly bending their bodies into a characteristic figure of eight shape. They live near the bottom and amongst the weed and feed on decaying vegetation. As they reach maturity, the larvae shed their skins to emerge as pupae.

The midge's pupal stage only lasts for a matter of days but is of great importance to trout and fisherman alike. The pupae have similar size and coloration ranges to those of the larvae. Generally speaking, their heads and thoraces are dark and constitute about a quarter of their overall body lengths. Their abdomens are clearly segmented and, when at rest, often take on a most convenient hooked shape. Four tufts of white gill filaments protrude from the fronts of their thoraces and they have similarly feathery, but very short, white tails which are believed to be used for absorbing dissolved salts from the water. As they mature, wing cases develop over their thoraces.

Midge pupae hover about close to the lake bed until ready to hatch and then rise slowly to the surface in a somewhat hesitant progression. Once there, they swim along horizontally just below the surface film or hang motionless immediately beneath it. As they start to hatch, their thoraces actually in the surface film, their bodies go rigid and the pupal cases split down the backs from the front, allowing the winged adults to emerge. The pupae have more difficulty in penetrating the surface film on still, calm days than they do when a breeze or wind causes ripples or waves which reduce the surface tension. This is significant when we come to consider fishing techniques.

When they appear, the adult midges are all similar in shape but vary enormously in size and colour, just as the larvae and pupae do. The midges, both male and female, have sharply segmented bodies and transparent, heavily veined wings which lie flat along the tops of their abdomens when at rest. Each has six legs which are relatively long for its body size and the males are identifiable by the two plumes which protrude from their heads. Midges mate on the wing and are often seen in dense clouds in the evenings, particularly along hedgerows and on the downwind side of tall trees or buildings. Once mated, the females return to the water to lay their eggs, their bodies taking on a familiar 'hanging hook' shape.

Artificial patterns

Although midge larvae are eaten in large quantities by trout, their habitat amongst the weed and silt of the lake bed and their extraordinary lashing motion make them very difficult to represent with artificial patterns. In addition, these creatures take so long to cover any distance that the angler's retrieve must be effectively non-existent. Several eminent anglers, Arthur Cove and Brian Clarke amongst them, have designed imaginative and realistic patterns to represent the larvae. However, the only one that I have used — and therefore, the only one that I feel qualified to quote — is a dressing by John Goddard which may be used to suggest either the red or the green larvae:

Hook	8–12 long-shank, down-eyed
Tying silk	Brown
Tail	Piece from the curly section of a red ibis quill or dyed green heron quill — to match body materials
Body	Crimson or olive condor herl covered with fluorescent floss of the same colour and ribbed with narrow silver tinsel
Thorax	Buff condor herl

In angling terms, the midge pupa differs dramatically from the larva. Trout may take the pupae whenever they are available but seem particularly susceptible when the insects are rising through the water or actually hatching. Countless dressings have been devised to represent these creatures but the following versions, all quite realistic and easy to tie, should meet most needs. They are so common-place that their origins seem impossible to identify:

Hook	10–14 (or even 16) down or straight-eyed
Silk	Black
Tail	A small bunch of white cock's hackle fibres, trimmed to an eighth of an inch at most
Body	*Either:* Black floss silk ribbed with stripped white cock's hackle stalk; or red floss silk ribbed with fine or medium silver tinsel (depending on hook size); or green floss silk ribbed with fine gold tinsel; or brown floss silk ribbed with fine gold wire
Thorax	Bronze peacock herl for the black and red versions; dubbed mole's fur for the brown and green ones
Breathing filaments	White wool, clipped short and fluffed out

Notes:
1. In all midge pupae the body should be carried well round the bend of the hook.
2. The gaps between the turns of tinsel ribbing on the red midge pupa should be approximately the same width as the tinsel itself.

3. To suggest the orange silver pupae which are in evidence during the early months of the season, a pad of hot orange floss silk may be tied in over the thorax of a red pupa to represent the natural's wing cases.
4. If very small hooks are employed, white cock's hackle fibres should be used instead of wool for the gill filaments, and they may be angled backwards over the thorax further to reduce the dressing's overall size.

Those who frequently encounter the Phantom midge should find that the following pattern will serve when the pupae are rising to hatch. The natural, which has no breathing filaments on its head and which has two small, paddle-like gills where its tail should have been, turns from a pale, translucent greenish yellow to a rusty brown colour before hatching. It is most commonly in evidence from late July to the end of the season.

Hook	14–16 straight-eyed
Tying silk	Brown
Abdomen	Pale yellow floss silk ribbed with fine silver tinsel
Thorax	Rusty pheasant tail fibres wound into a small lump just behind the eye of the hook

Trout rarely seem to feed on adult, winged midges and, for this reason, most fishermen tend to concentrate on the pupal stage of the insect's life. Although several artificials have been designed specifically to imitate adult midges, I must confess that I have never used one myself. However, I have had some considerable success with a small hackled Blue Upright when the trout have seemed to be feeding on the winged insects. The dressing is as follows:

Hook	14 or 16 up-eyed

Tying silk	Purple
Whisks	A small bunch of medium blue dun cock's hackle fibres, or none
Body	Stripped natural peacock herl
Hackle	Two medium blue dun cock's hackles, very short in the fibre

Presentation

The midge larva should be fished on the bottom with a sinking line and with virtually no movement at all. An occasional twitch of the line will impart some movement to the ibis fibre tail and may attract the attention of passing trout. The pattern may be used throughout the season but is likely to be easier to fish earlier on, before the weed has grown up.

There are several methods of fishing midge pupae but two of the most effective, described below, should suffice; the remainder are largely variations on themes. On warm, still days, when the naturals paddle about just below the surface desperately searching for a weak point in the film, and the trout take them with eager confidence, a single pupa may be used on a long, well greased leader. The fly should be allowed to hang static as close to the surface film as possible. If cruising trout can be seen it should be cast in their paths; if not, it should be left absolutely motionless for as long as possible. The take when it comes is, for me, one of the most dramatic experiences of stillwater fishing. There is rarely a hump in the water unless the fly is taken almost as soon as it has been cast. Instead, the leader simply draws slowly as the fish sucks in the pupa and proceeds unhurriedly on his way. The hook should be set instantaneously. When a trout takes an artificial presented thus, he has been truly deceived by the fly-dresser's sleight of hand. As no movement has been imparted to the pattern, which has been hung beneath the surface for the fish to inspect at leisure, it cannot have simply aroused the trout's curiosity or aggression.

The second technique is more suited to use when surface waves or a ripple make it more difficult for the fisherman to use or see a static, floating leader. A bushy, buoyant dry fly — generally a sedge pattern — is attached to the point of the leader and two or more pupae are tied on as droppers. The whole leader is greased to within an inch of each of the flies. The sedge ensures that the leader stays on the surface and can act as a float to indicate takes. The team of flies is cast from the bank or from an anchored boat and retrieved slowly and steadily.

As an alternative to this second technique, the sedge may be attached as the top dropper with midge pupae on the other dropper(s) and the point of an ungreased leader.

This allows the pupae to be fished at various depths and is particularly useful when no fish are moving on or near the surface.

The artificial adult midge may be useful either during a hatch or when the females return to the water to lay their eggs. In the former case, the pattern should be fished on a greased leader and some very slight animation may be given by occasional twitches of the line. In the latter, the pattern should be allowed to float motionless on the surface. In either case, the fly should be cast into the paths of cruising fish if possible. Otherwise it should be cast into specific areas in which fish are rising, very often small 'bays' or gaps in the weed.

The Pond Olive (*Clöeon dipterum*)

Of the various upwinged or day-flies to be found on stillwaters, the Pond Olive is the most widely distributed and probably the most important to the fisherman. It is to be found on small waters throughout England, Scotland

and Wales but is particularly common in the southern half of England. Hatches of Pond Olives occur from early May until the close of the season but seem to be most prolific during June, July and early August.

Life history

While it is an unusually tolerant creature, being able to survive under a wide range of water temperatures and conditions, the Pond Olive nymph does best in a reasonably warm environment. For this reason, it flourishes in smaller stillwaters, say up to 2–3 acres, particularly if they are sheltered from the wind. It tends to be less common in larger lakes where it usually lives in the relatively warm shallows. Very similar to the Lake Olive nymph in appearance, the Pond Olive nymph is generally about $\frac{1}{2}$ in. long overall although larger specimens are quite frequently found. While there is a wide range of colour variation between individuals, most of them have brown thoraces, darker brown wing cases and mottled yellow ochre and dark brown abdomens; some of these nymphs have greenish flecks on them and others have an overall olive tinge. Like its Lake Olive counterpart, the Pond Olive nymph has six sturdy legs, seven pairs of gills protruding from its abdomen and three tails, each light coloured and banded with brown.

Pond Olive nymphs are fast movers and can dart about amongst the weed in which they live with remarkable agility. They swim with a rapid, undulating movement of their bodies. When weed is sparse, they will be found amongst the leaves and decaying vegetation on the lake bed.

When ready to hatch, a nymph swims to the surface, its case expands and splits and the dun emerges to fly off almost immediately. This final process happens so rapidly that trout rarely have time to take the newly hatched duns

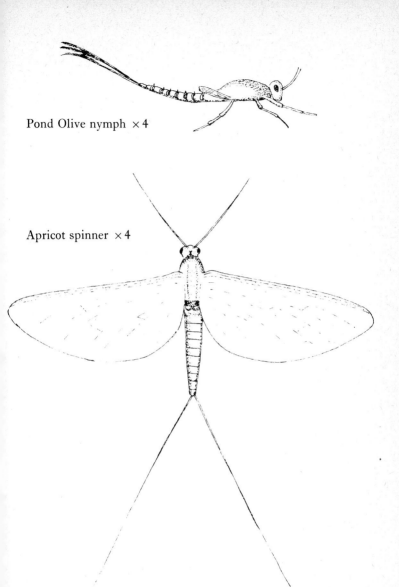

Pond Olive nymph × 4

Apricot spinner × 4

65

and they therefore generally concentrate on the hatching nymphs. The hatch, which usually takes place around the middle of the day, seems to be denser in warm, calm weather than when it is cold or blustery.

Pond Olive duns vary both in size and colour with the time of year. The early arrivals may be as much as $\frac{1}{2}$ in. long in the body; later ones are noticeably smaller. In general terms, the duns have dark olive-brown thoraces, dark olive-grey abdomens (usually with an orange tinge to them, especially towards the rear) and their single pairs of wings are a dark blue-grey. Their legs are significant as, being dull yellow at their extremities and almost black closer to the body, they bear a marked resemblance to the fibres of the furnace hackle used in a Greenwell's Glory. The adult Pond Olive has only two tails.

The Pond Olive spinner — the 'Apricot Spinner' — is a very much more elegant creature than it was in its previous incarnation as a dun. The female, for it is she who eventually arrives on the water as a snack for a trout, has a medium olive-green thorax but her abdomen is a brilliant apricot colour with amber markings. This gleaming spinner has completely transparent wings tinged with orange along their leading edges, two tails and her legs are a pale pastel green. Her husband is less impressive with a rather drab, pale grey body. The female spinners rarely return to the water before dusk but when they do so, they often arrive in vast numbers to lay their eggs before dying, spent, in the surface film.

Artificial patterns

Trout take the Pond Olive in all its stages — nymph, dun and spinner. The dun, which is only briefly on the water, is less important than the others. It is worth having four patterns to imitate this creature, the most widely distributed of all the upwinged stillwater flies.

David Collyer's American Gold Ribbed Hare's Ear Nymph, weighted and tied on a size 10 hook, is a most effective pattern with which to represent the Pond Olive nymph, just as it was with the Lake Olive one. The dressing is given on page 47.

As trout more often feed on the hatching nymph than on the newly emerged dun, a Gold Ribbed Hare's Ear is a useful fly when duns are leaving the water. In this case, I like to use one dressed with a furnace cock's hackle instead of the more usual pattern in which a few strands of fur are picked out from the thorax:

Hook	12–14, fine wire, up-eyed
Tying silk	Yellow
Whisks	Three or four guard hairs from a hare or rabbit's body fur
Body	Medium brown fur from a hare's ear, lightly dubbed and ribbed with oval gold tinsel
Hackle	Two or three turns of furnace cock's hackle

For the dun itself, it would be difficult to improve on a dry Greenwell's Glory. Nobody seems to know what Canon Greenwell intended this pattern to represent but its natural counterpart must have looked remarkably like a Pond Olive:

Hook	12–14 fine wire, up-eyed
Tying silk	Yellow: well waxed so that it takes on a slightly olive tint
Whisks	A small bunch of furnace cock's hackle fibres
Body	Tying silk, ribbed with fine gold wire
Hackle	Two bright, furnace cock's hackles

Once again, I believe that the omission of quill fibre

wings from this pattern must make it more realistic from the trout's point of view.

When Pond Olive spinners have been on the water I have had some considerable success with the following pattern, the origins of which I have been unable to establish.

Hook	14 fine wire, up-eyed
Tying silk	Olive
Tails	Half a dozen fibres from a pale grey speckled partridge's breast feather
Body	Stripped orange peacock herl ribbed with waxed scarlet tying silk
Hackle	Four or five turns of very pale blue dun cock's hackle at the shoulder with one turn of hot orange cock's hackle in front of them. Both hackles parted to stick out sideways from each side of the shoulder and held in position with a heavy figure of eight binding which represents the thorax

Presentation

The nymph can usually be fished at any stage of the season but seems to increase in effectiveness once the underwater vegetation has started to grow. It should be worked close to the bottom and around weed beds on as long a leader as necessary. The nymph should be retrieved in four to six inch jerks with pauses between them to allow it to sink again. As the pattern is often taken during the pauses between pulls, it is worth greasing the top few feet of the leader very lightly to make takes more visible. An interesting point when fishing this pattern is that fish have quite often come up from some considerable depth to take my nymph as soon as it has hit the water, particularly during the early days of the season. Why this should be so I cannot say; I have never noticed natural Pond or Lake Olives hatching at the time and a large enough proportion of the fish concerned have been wild brown trout to defeat the argument that this rather dumpy, tan coloured fly is taken for a pellet.

When a hatch is on and trout are rising, the Gold Ribbed Hare's Ear fished in, rather than on, the surface film can produce dramatic results. The leader should be greased to within about six inches of the fly and some life may be given to the artificial by very lightly twitching the line — the fly should not be moved across the water. If this fails, it is possible that the trout are taking the adult dun on the surface, and that a dry Greenwell's Glory may do the trick; it should be fished so that it floats high and without movement. If no trout are rising to the hatching insects, the nymph may be tried, fished just beneath the surface in a series of steady pulls with long pauses between them to allow it to sink back down again.

Pond Olive spinners rarely arrive on the water before nightfall but the angler who is able to get to the water very early in the morning, at or immediately after first light, may be able to take advantage of the trout's appetite for the spent insects. The leader should be lightly greased and the artificial should be cast in front of individual rising fish. So prolific are the falls of spinners on some occasions that great care will have to be taken to ensure that the artificial is selected by the trout from amongst the vast array of available naturals.

Sedges (*Trichoptera*)

After the midges, the sedges or caddis flies must rank as the most important group of insects from the stillwater fisherman's point of view and they can provide some of his most exciting sport. There are about 200 sedge species. Although there are marked differences of size and coloration between them, the diversities are not so pronounced as they are among the midges and it is quite practicable to use a fairly small range of patterns to represent the various species that the angler may expect to meet.

Life history

The larval stage is the first significant one in the sedge's life. Sedge (or caddis) larvae vary in length from $\frac{1}{4}$ in. to more than four times that size. The bodies of the larvae, segmented and slightly leathery in texture, generally taper from shoulder to tail. Each larva has a distinct, dark head and six quite powerful legs emanate from the front of its body. These creatures are bottom dwelling crawlers and almost all of them build protective cases around themselves. The cases, which are often to be found heaped up along the downwind shorelines of reservoirs and lakes, may be built of almost anything that comes to hand so long as the materials provide camouflage and a degree of physical protection. Common amongst the building materials are small pebbles, shells, sand, grit, twigs and leaves. Some larvae even cut lengths of hollow weed stem to use as houses, and day by day, they lumber about the bottom dragging their homes with them and retreating into them if danger threatens. They eventually pupate within their cases having sealed off the ends.

When, after a period of days or weeks, depending on the species, they emerge from their larval cases, sedge pupae vary in colour from light fawn through yellow and amber to pale green or dusky brown. With a range of sizes similar to that of the larvae, they are rather tubby creatures. The legs and antennae of the adult are visible, fully formed, beneath the pupal skin while the wings are generally encased outside it. One pair of legs, heavily

Sedge larva × 4

Sedge pupa × 4

Adult sedge × 4

fringed with hairs, is used to propel the pupa through the water. As these ungainly creatures make their way to the surface, they provide easy pickings for the trout which feed on them avidly.

In some species the pupae hatch into adult sedges as soon as they reach the surface. In others, they swim along just below the surface until they find something up which they can climb into the air.

Of the sedges which hatch out on the water, many take to the wing without a great deal of fuss as soon as they are able to do so, others scutter across the surface creating quite a commotion. Trout seem to delight in slashing at these struggling creatures, taking them with almost playful enthusiasm.

Adult sedge flies of interest to the fly fisherman vary in size from the Small Red sedge which is less than $\frac{1}{4}$ in. in length to the Great Red sedge at about an inch. They range in colour from dusty cream through an assortment of shades of brown to very dark grey. Many of them are mottled.

At rest, the sedges are readily identifiable by the roof-shaped arrangement of their wings over their backs. In flight, and this is how the angler most often sees them, they flutter along in a tail-down attitude and their wings, the hind ones markedly broader than the front ones, seem somehow to be disproportionately large. In addition to its four wings, each adult has six long, slender legs and a pair of antennae which vary in length according to species. The antennae are quite frequently as much as three times the insect's body length.

Artificial patterns

While several experimental patterns have been invented to imitate the sedges in their larval stages, few of them have been particularly successful. They have generally been made by sticking twigs, pebbles, shells or grains of sand to a suitably shaped underbody on a long-shanked hook. As sedge larvae move very slowly indeed when humping their cases about, any artificial must be fished correspondingly slowly along the bottom; all in all, a somewhat uninspiring process.

However, sedge pupae are an altogether different matter and artificials dressed to represent them can provide the stillwater fisherman with some of his best sport. Of the many patterns available, the one that follows has given me consistently good results. Whether it is original or not, I cannot say:

      Hook   10–12 or even 14 down-eyed

Tying silk	Brown
Underbody	Fine lead wire covered with white floss silk — or simply white floss silk
Body	Olive, yellow or fawn seal's fur, lightly dubbed and ribbed with fine gold tinsel
Legs/wing cases	One turn of partridge breast feather, tied as a beard
Head	Bronze peacock herl

The hatching sedge pupa is well represented by the Invicta, a 'traditional' pattern which now finds a place in the imitative fisherman's armoury:

Hook	10–14 down-eyed
Tying silk	Brown
Tail	Golden pheasant crest feather
Body	Yellow seal's fur, ribbed with fine gold oval tinsel
Body hackle	Red cock's, dressed 'Palmer'
Shoulder hackle	Red cock's with a few blue jay secondary fibres tied in as a false hackle
Wings	Hen pheasant tail feather slips (or mottled brown hen's quill slips which seem to be just as effective and are easier to handle)

There can be few better patterns to imitate the adult sedge on stillwaters than the deer hair dressing designed by John Goddard and Cliff Henry — the 'G and H Sedge'. It is not easy to dress but its great attraction is that it floats like a cork (until a fish takes it) and that it will put up with quite a lot of casting and retrieving. The dressing is as follows:

Hook	8–10 long shank
Tying silk	Green
Underbody	Dark green seal's fur dubbed onto silk and tied in at the bend; stretched beneath body and tied in at head once the deer hair has been spun on
Body	Deer hair spun onto hook shank and trimmed to sedge outline
Hackle	Two rusty dun cock's hackles wound slightly down body, the tops of the hackles trimmed off level with the top of the body, the hackle stalks left to represent the antennae

Notes:

Alan Tomkins, who tied the G and H Sedge illustrated amongst the photographs in this book, has pointed out that, although it looks more impressive, the deer hair should not be tied in too closely as doing so detracts from its floatability, and that one should beware of leaving too much deer hair below the hook shank, thus closing the gape.

For imitating smaller natural sedges, and for those who find deer hair patterns difficult to tie or uncomfortable to cast, Richard Walker has designed several first class sedge patterns. The one given here is taken from David Collyer's excellent book *Fly-Tying*, has caught me many good fish and is straightforward in construction:

Hook	10–12 up-eyed
Tying silk	Brown
Tag	Fluorescent orange wool
Body	Cock pheasant centre tail fibres
Wings	A bunch of red cock's hackles trimmed square at the tips, or a bunch of pheasant

| | tail fibres |
| *Hackle* | Two red cock's hackles |

Presentation

The various sedge patterns can usefully be fished from the beginning of June until the end of the season. The Invicta and the dry sedges will almost always be most successful during the evening when the naturals are hatching but the pupa should catch fish throughout the day, particularly during June, July and August.

In the daytime, the sedge pupa is best fished with a floating line and a long, very lightly greased leader. It should be worked close to the bottom with a slow, steady retrieve. It is often taken as it sinks through the water and, even when a trout intercepts it while it is on the move having reached its fishing depth, the take is often so gentle as to be almost imperceptible. This style of fishing requires a high level of concentration.

The Invicta, fished to represent a hatching sedge, should be worked very slowly, just beneath the surface, with a floating line and an ungreased leader. It should, if possible, be cast to moving fish. In contrast to the pupa, the takes are often quite savage and the fly is sometimes taken actually as it hits the water.

The floating pattern may be allowed to lie static on the water, or may be retrieved steadily, but sometimes produces more takes and more excitement if it is cast out and then brought back in a series of reasonably fast, wake-making pulls. This movement simulates the newly hatched adult sedge's attempts to get airborne. The trout tend to slash at a sedge pattern presented in this way, and miss it as often as not. It is, however, a rewarding and exciting technique, particularly at sunset when the naturals are in the air and the trout have thrown discretion to the wind.

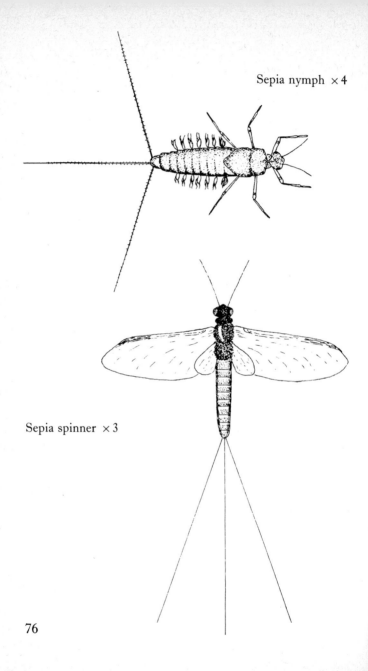

Sepia nymph × 4

Sepia spinner × 3

The Sepia dun (*Leptophlebia marginata*)

The Sepia dun is widely distributed throughout the British Isles but, seeming to show a slight preference for acid, peaty water, tends to be less common in the Midlands and Southern England than elsewhere. Where it is prolific, its appearance is doubly welcome to the angler for, following hard on the heels of opening day in many places, it is available to the trout when relatively little else is.

Life history

The Sepia nymph is a distinctive creature. Just over half an inch long, its sturdy, slightly flattened body is a very dark brown indeed and its wing cases are almost black. It has a heavy fringe of long, pointed mid-brown gill filaments down either side of its abdomen and its three tails, each as long as its body, are widely spread.

Although the Sepia nymph can swim, it does so badly and seems to spend most of its time crawling about amongst the decaying vegetation on the lake bed. For the greater part of its life, it stays in relatively deep water, only moving into the shallows (and particularly into heavily reeded areas) when ready to hatch. At some time between early April and early May, a build-up of gas beneath the nymphal case assists the insect on its journey to the surface and helps, perhaps even causes, the splitting of its skin and the emergence of the winged adult.

As its name suggests, the Sepia dun is essentially brown in colour. Its body, about $\frac{1}{2}$ in. long, is a dark sepia brown on top and a rather lighter shade below with even paler bands where the segments join. Its four wings are all a delicate, watery brownish yellow with heavy brown veining. It has three tails, widely spread, and its legs are mid-brown with a greenish tinge to them.

When the spinner appears, its body and legs are very

similar in size and coloration to those of the dun. Its glossy, colourless, transparent wings are veined in light brown and the forward ones have a grey streak towards the tips of their leading edges. The insects' tails are more than half as long again as their bodies.

Artificial patterns
C. F. Walker (*Lake Flies and Their Imitation*) notes that as the nymph rises to the surface, the layer of gas beneath the skin imparts a greyish, almost silvery sheen to it. For this reason there would seem to be some advantage in including a silver tinsel rib in an artificial intended to represent the mature nymph. The following simple nymph pattern has served me faithfully when either Sepia or Claret nymphs have been in evidence:

Hook	12–14 down-eyed
Tying silk	Black
Underbody	Fine copper wire, or none
Abdomen	Dark brown seal's fur ribbed with fine silver tinsel
Thorax	Dark brown seal's fur
Wing cases	Any dark brown or black quill fibre
Hackle	One turn of black hen's hackle — or none

Personally, I like to dress this pattern by building up the thorax with fine copper wire before winding on the dubbed seal's fur. It is easier to fish a slightly weighted pattern close to the surface when necessary than it is to persuade an unweighted pattern to sink.

A single dry fly, the Pheasant Tail, may be used to represent both Sepia dun and spinner. There are several variations on the theme but G. E. M. Skues's pattern (modified with a fine copper wire rib for strength) is an excellent one:

Hook	14 fine wire, up-eyed
Tying silk	Orange
Whisks	Three or four strands of honey dun cock spade feather
Body	Two or three strands of cock pheasant centre tail feather fibre, ribbed with very fine copper wire
Hackle	Brilliant rusty dun cock's hackle

Presentation

The Sepia nymph may realistically be fished from the beginning of April onwards. Until the duns start to appear, it should be worked slowly along the bottom on an ungreased leader. Once the first winged adults have hatched, it can be fished with a 'sink and draw' retrieve. It is often particularly successful in shallow water, especially the areas around reed beds and so on, and the margins should always be fished carefully before the water further from the bank is tried.

During a hatch of Sepia duns, which usually takes place around midday or in the early afternoon, the Pheasant Tail may be fished dry on a floating line and a leader greased to within six inches of the fly. The pattern should be cast to moving fish and allowed to sit motionless on the surface. Sadly, this technique is not always as successful as might be hoped; at this early stage in the season the trout are often lurking near the bottom and may be reluctant to take the surface food.

For some reason, Sepia spinners rarely fall on the water in any great numbers. While the fish will take them when a worthwhile opportunity presents itself, they form a relatively insignificant part of the average trout's diet. Should they be in evidence, the Pheasant Tail may be used as it is during a hatch of duns.

Shrimps (*Gammarus spp.*)

Although freshwater shrimps are prolific in most still-waters, trout do not seem to feed on them as avidly as they do, for example, on midge or sedge pupae. Nevertheless, they form a small but significant part of the fishes' diets and are valuable to the angler because they are available throughout the year. In fact, the number of trout taken on artificial shrimps in stillwaters greatly exceeds in proportion the number of shrimps found in autopsies. The reason for this only becomes apparent when we examine the various shrimp patterns used by fishermen and discover that they are tied in the most extraordinary range of colours — red, yellow, green, orange; who knows, a tartan one may even appear if the technique for dressing it can be perfected! It is clear that these rainbow hued patterns must be taken for something other than shrimps and I have reason to suspect that green and yellow dressings unwittingly play the parts of sedge pupae, especially in the summer months, and that the others simply arouse curiosity in trout. Be that as it may, a

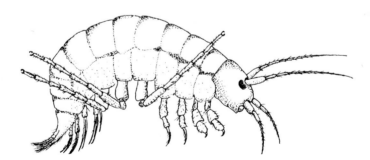

Freshwater shrimp × 4

realistic artificial shrimp is a useful addition to the armoury for use when there is little indication as to what the fish are feeding on.

The naturals

There are nine species of freshwater shrimp, the two commonest being *Gammarus pulex* and *Gammarus lacustris*. They vary from about $\frac{1}{4}$ in. to almost 1 in. in length and their armour plated humped backs are steeply rounded when the creatures are at rest. The shrimp has a complex mass of legs, the six walking ones being markedly longer than the others. Two sets of antennae emanate from its head, the longer pair being about half as long as the shrimp's body. Throughout most of the year the shrimp is a pale, watery, almost translucent, warm, fawnish colour or a dull yellow ochre. As the mating season approaches in mid-summer, it takes on a more brownish hue and dull red patches may develop on its back.

Freshwater shrimps, which are as commonly found in streams as in lakes, seem to dislike a stagnant, peaty or de-oxygenated environment and, in stillwaters, are most frequently to be found in and around weed beds near the inflows and outflows of streams. Most of them swim on their sides but this is no hard and fast rule and they will often as cheerfully potter along upright or upside down, apparently depending on their surroundings. At rest the shrimp adopts a curved posture, often wrapping itself around a leaf or a weed stem. As it starts to move, it straightens its body and propels itself almost entirely with a rapid, rhythmic movement of its legs. Shrimps are remarkably fast swimmers and have none of the jerky movement associated with other similar creatures. Those who have likened the freshwater shrimp to its saltwater counterpart may have misled people, for there is little resemblance between the two either in appearance or in movement.

Artificial pattern

There are many freshwater shrimp patterns available. Colonel Joscelyn Lane, John Goddard, C. F. Walker and Richard Walker have all produced excellent dressings and it would be difficult to say that one was better than another. The dressing given here has served me well and is, I believe, quite realistic:

Hook	10–12, down-eyed
Tying silk	Fawn
Underbody	Fine lead wire
Body	A mixture of light grey and a little light brown seal's fur or, for summer use, grey and red seal's fur; ribbed with fine gold wire
Legs	Iron Blue dun cock's hackle wound Palmer style up the body, trimmed off short along the sides and back and to a quarter of an inch beneath the body
Back	Stretched polythene strip

Presentation

An artificial shrimp may be used at any time of year. It should be fished close to the bottom, around weed beds and near the inflows and outflows of streams. It is best used on a floating line and as it is quite often taken while sinking through the water, it is generally worth very lightly greasing the top two-thirds of the leader to make these offers more visible. The pattern should be retrieved in a series of steady, eight to twelve inch pulls with pauses between them to allow it to sink again.

Snails

Any angler might be forgiven for believing that of all the creatures that go to make up a trout's diet, snails would be the least rewarding in terms of imitative representation or suggestion. In fact, this is quite untrue and there is considerable scope for experimentation in this field. Slow moving, succulent and prolific snails provide easy pickings for trout in many places, particularly during the mid to late summer. Contrary to some old wives' tales, the gastropods' shells do not seem to be crushed in the fishes' stomachs. I have caught trout in the late summer which felt like well-filled bean bags, positively grating when grasped firmly and yielding a tight packed mass of snails in the subsequent autopsy.

The naturals

Of some 38 species of freshwater snails recorded in the British Isles, only 16 are commonly found in stillwaters. Leaving the two species of freshwater limpets aside as being of little interest to fish or fishermen, the remaining species may reasonably be divided into two groups for angling purposes — the Ramshorns and the Bladder snails. It should be said that both descriptions, particularly the latter one, are biologically inaccurate as each refers colloquially to a small and specific group of snails within an overall type. Nevertheless, they will suit our needs.

The Ramshorns have flattish, disc-like shells which are, in fact, tapered tubes, narrow at the centre and expanding in successive whorls to their apertures. They range from just over $\frac{1}{8}$ in. to 1 in. or slightly more in diameter, the commonest of them being about $\frac{1}{4}$ in. across. In colour, the Ramshorns vary from a dull, chalky off-white through several shades of brownish olive to almost black.

The Bladder snails are very much more solid looking

Top: Ramshorn snail

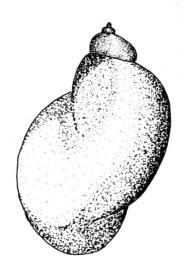

Right: Bladder snail

than the Ramshorns and resemble the common garden snail in general appearance. Their shells have fairly large apertures in bulbous bases and spiral to a pointed apex. The smallest of them are less than $\frac{1}{8}$ in. long and the largest rarely measure more than $\frac{3}{4}$ in. from aperture lip to point. Most of the Bladder snails have brown or red-brown shells of one shade or another but a few have rings or patterns on them and some are so dark as to be almost black.

Both types of snail live on or near the bottom, in and around weed beds, or on vertical surfaces, particularly wooden ones which accumulate algal growth. They are entirely vegetarian and mainly feed on algae although some of them can consume appreciable quantities of weed stem. For some reason as yet not fully understood, snails often float to the surface *en masse* in hot weather during

July, August and early September. Here they hang with their feet in the surface film and their shells suspended beneath it. Although snails are found in autopsies throughout the year, trout frequently gorge themselves when the creatures rise to the surface.

Artificial patterns

Published dressings designed to imitate snails are few and far between. Donald Downs has produced a remarkably realistic Ramshorn from a length of braided shoelace coiled into a disc on a small piece of balsa wood and secured to a hook. It is too early to say whether this is a particularly effective pattern but fishermen with an innovative bent may wish to experiment with it.

The following pattern, which was developed from one originally designed by Cliff Henry, has caught me a number of trout and proved particularly effective on a small stillwater in Gloucestershire during the very hot summer of 1976:

Hook	10 down-eyed, wide gape
Tying silk	Brown or black
Underbody	A flat-topped cone of cork, partly split and bound over the shank of the hook; the flat section facing the eye and coated with black varnish
Body covering	A half-width length of dark brown raffine ribbed, when dry, with fine copper wire. Two turns of bronze peacock herl wound over wet, clear varnish around the rim of the broad end

In addition to the artificial described above, I have taken many trout on the Black and Peacock Spider when the snail have been 'up'. It should be said at this juncture that this ubiquitous fly, first publicised by Tom Ivens in *Still Water Fly Fishing*, is a remarkable general pattern for use throughout the season. With it, I have had a limit in forty minutes at Rutland Water and many, many fish from both small stillwaters and rough streams. The dressing that I use is as follows:

Hook	10–12 down-eyed
Tying silk	Black
Underbody	Claret floss silk
Body	Bronze peacock herl wound into a rope with the tying silk
Hackle	Two or three turns of black hen's hackle

Presentation

The floating snail pattern should be fished static in the manner of a dry fly when the snail are at the surface. With so buoyant a dressing, there is little risk of the leader pulling the fly under, so the nylon may be left ungreased for the last foot or so.

The Black and Peacock Spider can also be used when the snail are up and should be fished as slowly as possible just beneath the surface on an ungreased leader. Takes are often very confident (rather than savage) and the hook should be set with a steady application of pressure rather than a more positive strike.

Sundries

The selection of species for inclusion in a pocket guide of this sort must inevitably be a somewhat arbitrary business. Some people may be disappointed that a particular insect, common on a water that they fish, has been excluded. The criteria that I have used in choosing the various fauna have been that each should be reasonably widely distributed throughout these islands, that each should constitute an appreciable part of the trouts' diet where it occurs and that each should present a practical proposition to the imitative fly dresser. Amongst those creatures that have not been included, the Yellow May dun was deemed to be too localised, surface dwellers such as the Pondskater and the Water-Measurer seemed to be too insignificant to the trout while Daphnia and the Phantom larva, one minute, the other transparent, were felt to be impractical for fly tying purposes. However, a few species are such borderline cases that it seems sensible to include some notes on them here.

Ants (*Hymenoptera*)

Ants appear to have been granted a place, albeit a small one, in the annals of fly fishing folklore. Few authors who have written on the subject of trout flies have been able to resist the temptation to suggest that every self-respecting fisherman should carry an artificial ant pattern in his fly box. While we are assured that winged ants flighting during hot, particularly thundery, weather in the summer

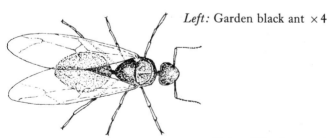

Left: Garden black ant × 4

Below: Wood ant × 4

months, do very occasionally fall onto the water in sufficient numbers to interest the fish. I have never witnessed such an event myself. Nor can any of my angling friends recount experiences of this phenomenon. I believe that a fall of ants is so rare an incident that the practical fisherman may reasonably ignore the possibility of encountering it. However, for those who may wish to arm themselves against an unlikely eventuality, the majority of ants are black, brown or rusty red and a variety of commercially dressed patterns is available to represent them. For myself, when I meet a fall I shall probably cut down an appropriately coloured dry fly of about the right size, fish it on the surface and see what happens.

Beetles (*Coleoptera*)

There is an enormous range of beetles, both aquatic and terrestrial, and a number of them are of interest to trout in one form or another.

Perhaps surprisingly, fish seem fairly indifferent to the adult forms of most aquatic beetles but will quite frequently take the larvae. These rather repulsive looking creatures vary from as little as $\frac{1}{4}$ in. to $1\frac{1}{2}$ in. or more in length. They generally have fairly bulky, segmented, peardrop shaped bodies, ranging in colour from a pale, translucent olive-grey to deep brown, and heads which look disproportionately small for their body sizes. Their six legs are necessarily sturdy and those that I have observed have spent most of their time crawling about on weeds and stones. So varied are aquatic beetles and their larvae that the angler who is interested in representing them would do well to take John Goddard's advice and tie patterns imitative of those most commonly found in his own water. I know of no commercial patterns available to suggest them.

In contrast, terrestrial beetles which are blown or fall

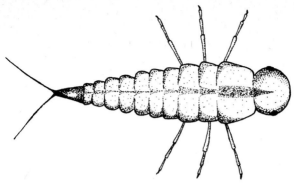

Water beetle larva × 4

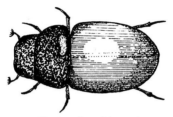

Coch-y-bonddu × 4

onto the water from time to time are often taken enthusiastically by trout. Amongst them, the Coch-y-bonddu and the Cockchafer are perhaps the best known, being prolific and widely distributed in Scotland and Wales, particularly in midsummer. Both are shiny black with red or brown wings and the traditional Coch-y-bonddu pattern, dressed on a size 10 or 12 hook and fished either dry or actually in the surface film, can produce good catches when the naturals are swarming. A similar sized Black and Peacock Spider (see 'Snails', page 85), dressed with a cock's hackle, a rather tubby body and, ideally, a silver tag to suggest a trapped air bubble, may profitably be used when any of the various black terrestrial beetles having dropped onto the water from marginal vegetation, and perhaps been blown clear of the bank, are in evidence. It should be fished dry with a greased leader and a floating line and may be twitched occasionally to suggest the laboured struggles of the natural trapped in the surface film.

The Blue Winged Olive (*Ephemerella ignita*)

Perhaps the best known of all the upwinged flies after the Mayfly, the Blue Winged Olive is primarily associated with running water. However, it is occasionally encountered on lakes and reservoirs and, when it occurs, can present the fisherman with splendid opportunities for classic dry-fly fishing. As its appearance on stillwater seems to be at best spasmodic and at worst rare, I do not normally carry patterns specifically designed to represent it but have enjoyed success using other dressings already in my fly box.

The nymph, about $\frac{1}{2}$ in. long, is similar in appearance to its Pond and Lake Olive counterparts and is adequately represented by David Collyer's American Gold Ribbed Hare's Ear nymph dressed on a size 10 hook. It should be fished slowly, close to the bottom, particularly around weed beds, in water up to eight or ten feet deep.

The dun, which hatches throughout the second half of the season, is distinctive in appearance. With an olive body about $\frac{1}{2}$ in. long and fairly dark blue-grey wings, it normally sits on the surface drying itself for some minutes after hatching. It may be adequately represented either with a Pheasant Tail dressed on a size 12 hook (see Claret dun, page 28) or with one of the several commercial patterns available. Skues's version of the Pheasant Tail spinner (see Sepia dun, page 78) may be used to suggest the Blue Winged Olive spinner and I have also had some success with a Claret spinner (see page 28) when spent naturals have been on the surface in quantity.

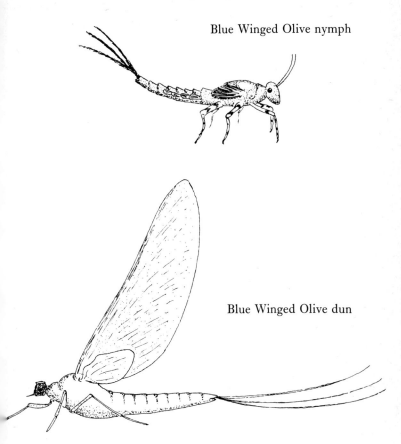

Blue Winged Olive nymph

Blue Winged Olive dun

Moths (*Lepidoptera*)

Although I have caught trout from both rivers and lakes on moth patterns, I have made no real study of the natural insects. John Goddard in *Trout Flies of Stillwater* devotes an interesting page and a half to these creatures and any fisherman wishing to delve deeper into the subject might do worse than to use his notes as a starting point.

The moths found at the waterside seem to range from ½ in. to 1 in. in length and most of them are fairly bulky creatures. They vary in colour from off white through cream to mottled brown and cream or slate grey and white. While most of them fold their wings across their backs in the familiar 'delta' configuration when at rest, they tend to spread-eagle themselves when trapped in the surface film.

Several commercial patterns are available to represent moths blown onto the water and, in addition to these, I have had some success with a fully dressed, dry, hackled March Brown. As moths generally appear at dusk, an artificial can usefully be fished in the late evening. It seems to be most successful when cast to individual fish moving close to the

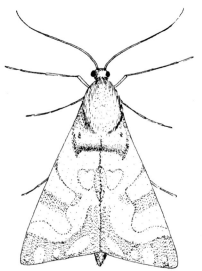

Brown China Mark moth × 4

bank, especially where the marginal vegetation is fairly dense. If a trout fails to take a pattern floating motionless on the water, it can sometimes be induced to do so by retrieving the fly in a series of quick twitches to emulate the struggles of the trapped natural.

Stoneflies (*Plecoptera*)

The Stoneflies have been relegated to the list of 'also rans' for several reasons. Only a few of the species within the order live in stillwaters, the majority being river dwellers. They are relatively localised in their distribution and are largely confined to cold, stony habitats in Scotland and the North Country. And, although trout undoubtedly consume both Stonefly nymphs and adults in substantial numbers, the former, while easy for the fly dresser to imitate, give the fisherman an outstandingly difficult task in terms of presentation.

At this juncture I should confess that I have never used any Stonefly pattern myself, so what follows must inevitably be drawn from the experience of those who know these insects better than I do.

The nymphs which, as the name suggests, live among the rocks and stones of the lake bed, are not unlike the nymphs of upwinged flies in general appearance. They differ in that they have only two tails each and are generally rather larger than their Ephemerid counterparts. The creatures' sturdy legs show them to be strong crawlers.

When mature, the nymph makes its way to dry land where the adult hatches. Most Stoneflies are brown although two or three species are a yellowish colour and all have four glossy, heavily veined wings which are folded flat and narrow over their abdomens when at rest. In many species, the wings of the male are no more than a gesture, being so small as to be useless. The insects range from rather less than $\frac{1}{4}$ in. to over 1 in. in length and each has

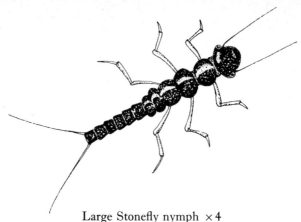

Large Stonefly nymph ×4

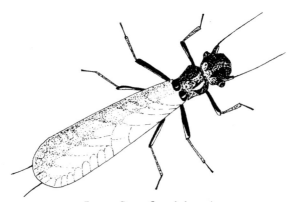

Large Stonefly adult ×4

a fairly long pair of antennae and two tails. Even in their winged state these creatures are noticeably more at home on land than in the air and, to those who fish northern lochs and lakes, the sight of the winged adults scuttling about on and among the shoreline boulders will be a

94

familiar one. The females, returning to the water to lay their eggs, present patrolling trout with tempting opportunities.

Many dressings have been designed to represent Stone-fly nymphs and winged adults, and several of them seem appropriate to stillwaters. The best course for an angler unfamiliar with an area in which the naturals are commonly found must be to seek the advice of a knowledgeable local fisherman or tackle dealer.

Bibliography

Books

Burton, John. *The Oxford Book of Insects*. Oxford University Press, 1968.

Clarke, Brian. *The Pursuit of Stillwater Trout*. A. & C. Black Ltd., 1975.

Collyer, David J. *Fly-Dressing*. David and Charles (Holdings) Ltd., 1975.

Goddard, John. *Trout Fly Recognition*. A. & C. Black Ltd., 1966.

Goddard, John. *Trout Flies of Stillwater*. A. & C. Black Ltd., 1969.

Goddard, John. *The Super Flies of Stillwater*. Ernest Benn Ltd., 1977.

Harris, J. R. *An Angler's Entomology*. Collins, 1952.

Ivens, T. C. *Still Water Fly Fishing*. Andre Deutsch Ltd., 1952.

Lane, Col. Joscelyn. *Lake and Loch Fishing*. Seeley Service and Co. Ltd.

Lawrie, W. H. *Modern Trout Flies*. Macdonald, 1972.

Veniard, John. *Fly Dressers' Guide*. A. & C. Black Ltd., 1970.

Veniard, John. *Further Guide to Fly Dressing*. A. & C. Black Ltd., 1972.

Walker, C. F. *Lake Flies and Their Imitation*. Herbert Jenkins, 1969.

Williams, A. Courtney. *A Dictionary of Trout Flies*. A. & C. Black Ltd., 1973.

Freshwater Biological Association Scientific Publications

No. 13 'A Key to the British Fresh- and Brackish-water Gastropods', by T. T. Macan, 1977.

No. 15 'A Revised Key to the Adults of the British Species of Ephemeroptera', by D. E. Kimmins, 1972.

No. 16 'A Revised Key to the British Water Bugs (Hemiptera —Heteroptera)', by T. T. Macan, 1965.

No. 20 'A Key to the Nymphs of British Species of Ephemeroptera', by T. T. Macan, 1970.

No. 24 'A Key to the British Species of Simuliidae (Diptera) in the Larval, Pupal and Adult Stages', by Lewis Davies, 1968.

No. 32   'A Key to British Freshwater Crustacea: Malaco-straca', by T. Gledhill, 1976.

No. 35   'A Key to the Larvae and Adults of British Freshwater Megaloptera and Neuroptera', by J. M. Elliott, 1977.

Upwinged fly identification

Month	Body length	No. of wings	No. of tails	Wing colour	Body colour	Insect
April	12–14 mm	4	3	Pale fawn	Dark sepia brown	Sepia dun
May and June	10–12 mm	2	2	Grey	Olive (tinged grey or brownish)	Pond Olive dun
	11–13 mm	2	2	Pale blue-grey	Red-brown	Lake Olive dun
	4–6 mm	2	3	Off white	Cream	Caenis
	11–13 mm	4	3	Very dark grey	Very dark brown	Claret dun
	14–17 mm	4	3	Heavily veined; brown or black on grey	Creamy white with brown markings	Mayfly
	12–14 mm	4	3	Pale fawn	Dark sepia brown	Sepia dun
July	10–12 mm	2	2	Grey	Olive (tinged grey or brownish)	Pond Olive dun
	4–6 mm	2	3	Off white	Cream	Caenis
	11–13 mm	4	3	Very dark grey	Very dark brown	Claret dun
	12–14 mm	4	3	Dark blue-grey (raked back when at rest)	Rusty brown	Blue Winged Olive
August	10–14 mm	2	2	Grey	Olive (tinged grey or brownish)	Pond Olive dun
	11–13 mm	2	2	Pale blue-grey	Red-brown	Lake Olive dun
	4–6 mm	2	3	Off white	Cream	Caenis
	12–14 mm	4	3	Dark blue-grey (raked back when at rest)	Rusty brown	Blue Winged Olive
September	10–12 mm	2	2	Grey	Olive (tinged grey or brownish)	Pond Olive dun
	11–13 mm	2	2	Pale blue-grey	Red-brown	Lake Olive dun
	12–14 mm	4	3	Dark blue-grey (raked back when at rest)	Rusty brown	Blue Winged Olive

The stillwater flyfisherman's calendar

Species	March	April	May	June	July	August	September	October
Alder larvae		●	●	●				
Caenis			n	●	●	●		
Claret dun			n ●	●	●			
Corixids	●	●	●	●	●	●	●	●
Crane flies						●	●	
Damsel nymphs	n	n	n	n	n	n		
Fry					●	●	●	
Lake Olives			n ●	●		●	●	●
Lice	●	●	●	●	●	●	●	●
Mayflies			●	●				
Green midges	Small	Small		Large	Large			
Large Red midges					●	●	●	●
Black midges	●	●			●	●	●	●
Small Brown midges				●	●	●	●	●
Orange-Silver midges		●	●	●				
Pond Olives		n	●	●	●	●	●	●
Sedges				●	●	●	●	●
Sepia duns		●	●					
Shrimps	●	●	●	●	●	●	●	●
Snails					●	●	●	
Ants						●		
Beetles			●	●	●	●	●	●
Blue Winged Olives					●	●	●	
Stoneflies	●	●	●	●	●	●	●	●

- - - Available as nymph ⎫
——— Available as adult ⎭ where applicable.

99